I0003360

Node.js: Novice to Ninja

Copyright © 2022 SitePoint Pty. Ltd.

- **Product Manager:** Simon Mackie
- **Technical Editor:** Ivaylo Gerchev
- **English Editor:** Ralph Mason
- **Cover Designer:** Alex Walker

Notice of Rights

All rights reserved. No part of this book may be reproduced, stored in a retrieval system or transmitted in any form or by any means, without the prior written permission of the publisher, except in the case of brief quotations embodied in critical articles or reviews.

Notice of Liability

The author and publisher have made every effort to ensure the accuracy of the information herein. However, the information contained in this book is sold without warranty, either express or implied. Neither the authors and SitePoint Pty. Ltd., nor its dealers or distributors will be held liable for any damages to be caused either directly or indirectly by the instructions contained in this book, or by the software or hardware products described herein.

Trademark Notice

Rather than indicating every occurrence of a trademarked name as such, this book uses the names only in an editorial fashion and to the benefit of the trademark owner with no intention of infringement of the trademark.

Published by SitePoint Pty. Ltd.
10-12 Gwynne St,
Richmond, VIC, 3121
Australia
Web: www.sitepoint.com
Email: books@sitepoint.com
ISBN 978-1-925836-52-3 (print)
ISBN 978-1-925836-53-0 (ebook)

Printed and bound in the United States of America

About Craig Buckler

Craig is a UK-based freelance full-stack web developer, writer, and speaker who's passionate about standards and performance.

He began coding in the 1980s and started client-side JavaScript development on its release in 1995 when DHTML, spacer GIFs, and marquees were considered sophisticated. You may have encountered his work at SitePoint, where he's written more than 1,200 tutorials, and books including *Jump Start Web Performance*[1], *Browser DevTool Secrets*[2], and *Docker for Web Developers*[3].

Craig used Node.js from the start and hopes this book is a great first step on your server-side JavaScript journey. Contact him on Twitter @craigbuckler or at craigbuckler.com.

About SitePoint

SitePoint specializes in publishing fun, practical, and easy-to-understand content for web professionals. Visit **https://www.sitepoint.com/** to access our blogs, books, newsletters, articles, and community forums. You'll find a stack of information on JavaScript, PHP, Ruby, mobile development, design, and more.

---

[1] https://www.sitepoint.com/premium/books/jump-start-web-performance/
[2] https://www.sitepoint.com/premium/books/browser-devtool-secrets/
[3] https://www.sitepoint.com/premium/books/docker-for-web-developers/

Table of Contents

Chapter 3: **Your First Node.js**

Application 18

Chapter 4: **How to Debug Node.js**

Scripts.. 38

Chapter 10:   **Using Database Storage** 189

Chapter 11:   Using WebSockets.................. 221

Preface

This book will help you get started with Node.js in the shortest possible time. Within a few days you should have enough knowledge to write simple applications.

Prerequisites

This course is for web developers taking their first steps with Nodes.js. Ideally, you should understand web development concepts and technologies:

- web browsers (client-side HTML, CSS, and JavaScript)
- web servers (code to serve web pages and APIs)

It will help if you already know some JavaScript—perhaps from writing client-side scripts. This book explains some aspects of JavaScript in relation to Node.js, but you won't find deep dives into variables, loops, functions, objects, and so on.

A little knowledge of the command line, Git, and code editors such as VS Code will also be useful.

Code Samples

Code in this book is displayed using a fixed-width font, like so:

```
<h1>A Perfect Summer's Day</h1>
<p>It was a lovely day for a walk in the park.
The birds were singing and the kids were all back at school.</p>
```

Where existing code is required for context, rather than repeat all of it, ⋮ will be displayed:

```
function animate() {
  ⋮
new_variable = "Hello";
}
```

Some lines of code should be entered on one line, but we've had to wrap them because of page constraints. An ↪ indicates a line break that exists for formatting purposes only, and should be ignored:

```
URL.open("https://www.sitepoint.com/responsive-web-
↪design-real-user-testing/?responsive1");
```

Tips, Notes, and Warnings

 ### Hey, You!

Tips provide helpful little pointers.

 ### Ahem, Excuse Me ...

Notes are useful asides that are related—but not critical—to the topic at hand. Think of them as extra tidbits of information.

 ### Make Sure You Always ...

... pay attention to these important points.

 ### Watch Out!

Warnings highlight any gotchas that are likely to trip you up along the way.

What is Node.js?

Chapter

1

Node.js is a JavaScript runtime, which means it runs programs written in JavaScript. Most developers use it to create command-line tools or web server applications.

 Skip Ahead?

That's everything you need know about Node.js. If you're eager to start programming, skip ahead to Chapter 2. That said, it's worth revisiting this chapter later to learn about Node's advantages and core features.

 JavaScript, JScript, ECMAScript, ES6, ES2015?

To make learning more confusing for beginners, JavaScript has many names. It started life as "Live Script" in 1994. Netscape rebranded it as "JavaScript" following a hasty deal with Sun Microsystems, despite it bearing little resemblance to Java or lightweight scripting. Microsoft couldn't use that name, so it became "JScript" in Internet Explorer.

In 2005, Mozilla (which grew out of Netscape) joined ECMA International and standardized the language as "ECMAScript"[1]. Versions 1 to 3 documented the evolution of JavaScript up until 1999. Version 4 was abandoned, but ECMAScript 5 became the standard supported by most browsers in 2009.

Work then started on ECMAScript 6—or "ES6". The final specification was approved in 2015, which led to yet another name: "ES2015". New specifications now arrive every year.

Rightly or wrongly, this course refers to "JavaScript" throughout. References to specific versions (such as ES9/ES2018) are only made if they affect the version of Node.js you need to use.

---

[1] https://www.ecma-international.org/publications-and-standards/standards/ecma-262/

Node.js was initially developed by Ryan Dahl. He took the V8 JavaScript engine from Google's Chrome browser, added some APIs, wrapped it in an event loop, and launched it as an open-source product on Linux and macOS in 2009. The Windows edition arrived in 2011.

The Node Package Manager (npm) was introduced in 2010. It allowed developers to use code modules published by others in their own projects. There was no official ECMAScript module standard at the time, so Node.js and npm adopted CommonJS.

The first (non-beta) release of Node.js arrived in 2015, with updates promised every six months.

Node.js wasn't the first JavaScript runtime, but unlike other options—such as Rhino[2] and SpiderMonkey[3]—its popularity grew exponentially. Even those writing PHP, Python, Ruby or other languages often use Node.js tools to supplement their development processes.

Why Learn Node.js?

JavaScript is the most-used language on GitHub[4], and it's >ranked highly by developers[5]. Companies including Netflix, Uber, Trello, PayPal, LinkedIn, eBay, NASA and Medium have adopted Node.js, and most professional developers will have encountered Node.js tools.

Below, we'll look at some of the reasons you should consider using Node.js.

It's JavaScript

JavaScript is used on trillions of web pages, where it has a browser monopoly. Every professional web developer requires JavaScript knowledge to program

---

[2.] https://github.com/mozilla/rhino
[3.] https://spidermonkey.dev/
[4.] https://madnight.github.io/githut/
[5.] https://insights.stackoverflow.com/survey/2021#most-popular-technologies-language

client-side applications.

Server-side languages are more diverse. Historically, developers could opt for PHP, Ruby, Python, C# (ASP.NET), Perl, or Java, but these have different syntaxes and concepts. It can be difficult to switch contexts, so larger project teams often split into frontend and backend developers.

Node.js allows a developer with frontend JavaScript knowledge to leverage their skills on the backend. It won't make you a full-stack developer overnight, but the concepts are similar, and there's less rigmarole when formatting JSON, handling character sets, using WebSockets, and so on.

 JavaScript Alternatives

> Some developers prefer languages such as TypeScript, PureScript, CoffeeScript, Reason, and Dart, which can transpile to JavaScript and run in a browser or Node.js. Ultimately, it still results in JavaScript code.

It's Fast

Most server-side languages are fast enough, but few match the speed of Node.js. The V8 engine is quick, and it evolves rapidly, having the weight of Google and Chrome behind its development. Node.js also has a non-blocking, event-driven I/O.

Let's go through that again with less jargon. Most languages use synchronous blocking execution. When you issue a command—such as fetching information from a database—that command will halt further processing and complete before the runtime progresses to the next statement. To ensure that multiple users can have access at the same time, web servers such as Apache create a new processing thread for every request. This is an expensive operation, and Apache has a default limit of 150 concurrent connections. Busy servers can become overloaded.

Node.js code (and browser JavaScript) runs on a single processing thread. Long-running tasks such as a database query are processed asynchronously, which doesn't halt execution. The task runs in the background, and Node.js continues to the next command. When the task is complete, the returned data is passed to a callback function. A Node.js program can have many hundreds of ongoing operations that are completed whenever they're finished, meaning that the processor is free to tackle other tasks.

Asynchronous programming has challenges, but it's possible to create fast Node.js applications that scale well.

It's Real-time

Web platform features such as WebSockets[6] and server-sent events[7] permit real-time functionality—such as instant data updates, live chat, multiplayer games, and more. These can be difficult to implement in traditional server-side languages, where they often require third-party services. Real-time functionality in Node.js is significantly easier.

It's Lightweight

The Node.js runtime is small and cross-platform. As well as catering for Linux, macOS, and Windows, you find editions for embedded systems, the Raspberry Pi, and even SpaceX rockets.

It's Modular

Node offers a minimal standard library with good documentation[8]. It contains basic functions for error handling, file system access, network operations, and logging.

For everything else …

---

6. https://developer.mozilla.org/Web/API/WebSockets_API
7. https://developer.mozilla.org/Web/API/Server-sent_events
8. https://nodejs.org/api/

It's Extendible

Node.js has the largest package registry in the world, with more than one million modules. You'll find pre-written code for task runners, loggers, database connectors, image processors, code compilers, web servers, API managers, and even client-side libraries.

The npm command-line tool is provided with Node.js and makes it easy to install, update, and remove modules. You can also use it to install global modules so Node.js scripts can run as commands from anywhere on your system.

It's Open Source

Node.js is an open-source project. The runtime is free to u page in your browser and you'll be directed to se without any commercial restrictions. The majority of modules are also free, because they're submitted by the community for the benefit of other developers.

It's Everywhere

This course concentrates on web applications, but you can use Node.js to create serverless functions, deployment scripts, cross-platform command-line tools, and even complex graphical apps such as VS Code, Slack, and Skype—all of which use the Electron framework[9].

As a web developer, you'll almost certainly encounter Node.js, even if it's not a core part of your technology stack. Knowing a little Node.js could help your projects and career. You'll have a better insight into the possibilities available to modern web applications.

---

[9]. https://www.electronjs.org/

What About Deno?

Node.js is a cross-platform, V8-based JavaScript runtime released by Ryan Dahl in 2009.

Deno[10] is a cross-platform, V8-based JavaScript runtime released by Ryan Dahl in 2020.

Deno smooths over some cracks and inconsistencies of Node.js, with the benefit of a decade's worth of hindsight. It directly supports TypeScript without a compiler, uses ES6 modules rather than CommonJS, replicates many browser APIs (`window` , `addEventListener` , `Fetch` , Web Workers, etc.), and provides built-in tools for linting, testing, and bundling.

Deno is great—but it's new, and yet to achieve a fraction of Node's popularity. The frameworks are similar: if you know one, it's easy to switch to the other.

Summary

In this chapter, you've learned that Node.js is a popular JavaScript runtime that's uniquely suited to web development. I've summarized it in this chapter's video[11]. Chapter 2 describes how to install Node.js on your platform of choice.

Quiz

Many chapters in this course end with a quick quiz to ensure you've grasped the concepts. Beware! Some questions are designed to catch you out, so make sure you've been paying attention! Answers can be found in Appendix A, at the back of the book.

1. What is Node.js?

---

10. https://deno.land/
11. https://spnt.co/nodevid1

a. A JavaScript runtime.
b. A tool for creating command-line, GUI, and web applications.
c. A cross-platform programming framework.
d. All of the above.

2. What is JavaScript's relationship to Java?

a. JavaScript is a cut-down version of Java.
b. JavaScript is a cross-platform version of Java.
c. JavaScript is Java that runs in a web browser.
d. JavaScript is a marketing name.

3. What is *not* another name for JavaScript?

a. ECMAScript
b. TypeScript
c. JScript
d. ES2015

4. What best describes the Node.js non-blocking, event-driven I/O?

a. Code that runs in separate processing threads.
b. Code that runs synchronously; the next command runs after the current command has completed.
c. Code that runs asynchronously; the next command could run before the current command has completed.
d. Code that runs in parallel with other processes.

5. What is npm short for?

a. Node Package Manager
b. Node Program Maintenance
c. Node Parsing Methods
d. Node.js Perfect Manual

Install Node.js

You won't get far on your Node.js journey without installing the runtime first! You have three primary options:

- Install Node.js on your local development machine running Linux, macOS, or Windows.

  This is the easiest choice, and the best way to get started—*and it's the option we'll be taking here.*

- Install Node.js via a virtual machine (typically Linux) running on Hypervisor software such as VMware[1], VirtualBox[2], Parallels[3], or Hyper-V[4].

  This won't affect your main OS, so you can experiment without risk.

  Windows users should also consider the Windows Subsystem for Linux 2 (WSL2)[5], which offers a highly integrated Linux environment. Follow the Linux instructions accordingly (found below in the "How to Install Node.js on Linux (or Windows WSL2)" section).

- Containerization software such as Docker[6].

  Docker provides a wrapper around applications known as a **container**. You'll use Docker in later chapters to install software such as databases, but you can also develop, debug, and deploy Node.js apps in a similar way.

  A configured container runs identically on every OS, so it's ideal when working in teams where members have different devices.

Node.js apps will work cross-platform, but there are differences in file systems and supported software. Web applications are typically deployed to a Linux

---

[1.] https://www.vmware.com/
[2.] https://www.virtualbox.org/
[3.] https://www.parallels.com/
[4.] https://docs.microsoft.com/virtualization/
[5.] https://www.sitepoint.com/wsl2/
[6.] https://www.docker.com/"

server, so developing on a Linux OS, virtual machine, or Docker container can help to avoid compatibility issues.

 Node Version Manager

Node Version Manager[7] (nvm) is a tool that allows multiple editions of Node.js to be installed on the same Linux, macOS, or Windows WSL system. This can be practical if you're working on two or more projects using different versions of Node.js.

Choosing a Node.js Version

Install a recent release of Node.js unless you're supporting a legacy application with specific requirements.

Even-numbered Node.js versions—such as 16, 18, and 20—focus on stability and security with long-term support (LTS). Updates are provided for at least two years, so I recommend them for live production servers. You should install an identical version on your development machine.

Odd-numbered versions—such as 15, 17, 19—are under active development and may have experimental features. They're fine for development if you're learning, experimenting, or upgrading frequently.

Node.js 16 was used to develop the example code in this course. However, Node.js generally has good backward compatibility, and applications written in earlier editions of the framework *usually* run in later versions.

How to Install Node.js on Linux (or Windows WSL2)

Open the nodejs.org[8] home page in your browser and you'll be directed to

---

7. https://github.com/nvm-sh/nvm
8. https://nodejs.org/

download an installation package appropriate for your OS. However, it's most practical to use the package manager built into your OS[9]. Ubuntu and Debian binaries are available from NodeSource[10] and, using version 16.x as an example, you can install Node.js from an Ubuntu bash terminal like so:

```
curl -fsSL https://deb.nodesource.com/setup_16.x | sudo -E bash -
sudo apt-get install -y nodejs
```

Verify that Node.js and npm are installed correctly by running the following commands in the terminal to view their version numbers:

```
node -v
npm -v
```

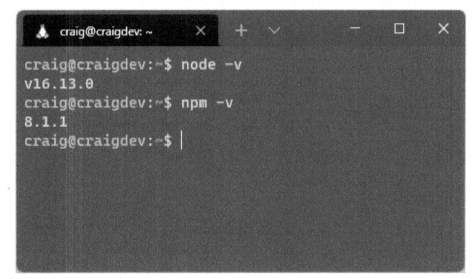

2-1. Node and npm versions

Configuring npm Global Permissions on Linux

The Node Package Manager command-line tool is provided with Node.js and makes it easy to install, update, and remove modules. Where practical, Node.js

---

9. https://nodejs.org/en/download/package-manager/

10. https://github.com/nodesource/distributions/blob/master/README.md

apps can be installed globally and run from anywhere—such as the Rollup[11] bundler for optimizing client-side JavaScript:

```
npm install rollup --global
```

This command results in a `EACCES permission error`, because you're not running as a superuser or using `sudo`. However, `sudo` grants unlimited permissions to global scripts. *I'd rather not run my own code that way, let alone someone else's!*

A better option is to change the default `npm` directory[12] to one owned by you. Create a new directory for global modules, then configure npm, like so:

```
mkdir ~/.npm-global
npm config set prefix '~/.npm-global'
```

Then, open `~/.bashrc` or `~/.profile` in an editor such as `nano`:

```
nano ~/.bashrc
```

Next, add the following lines to the end of the file:

```
export NPM_GLOBAL="$HOME/.npm-global"
export PATH="$NPM_GLOBAL/bin:$PATH"
```

Restart the Bash terminal or update the system manually with `source ~/.bashrc`.

You can now install global modules without `sudo`—including updates to npm itself:

---

11. https://rollupjs.org/
12. https://docs.npmjs.com/resolving-eacces-permissions-errors-when-installing-packages-globally

```
npm install npm --global
```

How to Install Node.js on macOS

Open the nodejs.org[13] home page in your browser and you'll be directed to download the Node.js `.pkg` installer for macOS. Launch the file, agree to the terms, and continue the installation.

Verify that Node.js and npm are installed correctly by running the following commands in the terminal to view their version numbers:

```
node -v
npm -v
```

How to Install Node.js on Windows

You can perform a Windows installation of Node.js in three ways:

- on Windows directly
- on a Linux distro installed in WSL2 (refer to the "How to Install Node.js on Linux (or Windows WSL2)" section above)
- on both Windows and Linux!

To install on Windows, open the nodejs.org[14] home page in your browser and you'll be directed to download the Node.js `.msi` installer. Launch the file, agree to the terms, and continue the installation.

Verify that Node.js and npm are installed correctly by running the following commands in the terminal to view their version numbers:

```
node -v
```

[13.] https://nodejs.org/
[14.] https://nodejs.org/

```
npm -v
```

How to Install Node.js on Other Devices

If you're using another device, chances are you'll find a Node.js distribution *somewhere*. It may not be on the standard nodejs.org website, so try Googling "install Node.js on [my-device's-name]".

For example, searching for "Install Node.js on Raspberry Pi" provides many resources[15] for installing Node.js on different editions of the hardware.

Run JavaScript Commands in the Node.js REPL

Node.js provides a read-evaluate-print loop (REPL) language environment. It will be familiar if you've ever opened a browser's developer tools console, and it's useful for testing snippets of code.

Start the REPL from your terminal by entering `node`. You'll see a prompt such as this:

```
Welcome to Node.js v16.12.0.
Type ".help" for more information.
>
```

Enter a JavaScript command or expression at the `>` prompt. For example:

```
> const myname = 'World';
```

(Replace "World" with your own name in quotes.)

You'll see *undefined* returned, because the expression doesn't output anything. Now enter the following, to see "Hello World" (or whatever name you used):

---

15. https://gist.github.com/stonehippo/f4ef8446226101e8bed3e07a58ea512a

```
> console.log(`Hello ${ myname }`);
```

Again, *undefined* is shown because *console.log()* outputs a string and doesn't return a value.

You can enter any JavaScript expression. It's not necessary to wrap it in a *console.log()* . For example:

```
$ node
Welcome to Node.js v16.12.0.
Type ".help" for more information.
> 2+2
4
> const myname = 'World'
undefined
> `Hello ${ myname }`
'Hello World'
>
```

Finally, press Ctrl | Cmd + D to exit the REPL console.

You're unlikely to use the REPL environment on a daily basis, but it can be useful for evaluating simple expressions before adding them to a script.

Summary

In this chapter, you've learned how to install Node.js on a variety of devices and run JavaScript commands in the REPL console. I've summarized it in this chapter's video[16]. In the next chapter, you'll write your first JavaScript-powered console and web applications.

Quiz

1. Versions of Node.js are available for:

a. most Linux distributions

[16.] https://spnt.co/nodevid2

b. macOS

c. Microsoft Windows

d. all of the above

2. What is nvm used for?

a. It's an alternative to the standard npm.

b. It can install and manage different versions of Node.js on one device.

c. It's a module search system.

d. It's a text editor specifically designed for JavaScript applications.

3. What is REPL short for?

a. read-evaluate-print loop

b. read-execute-print loop

c. run-evaluate-print loop

d. read-execute-primary loop

Your First Node.js Application

Chapter

3

In this chapter, you'll write, run, and debug your first Node.js programs. To keep it simple, these won't use any third-party modules or npm. They're self-contained scripts that use the standard library provided in Node.js[1].

Your First Console App

Command-line console applications can be useful for automating tasks, formatting data, manipulating files, or any other laborious job that's best handled by a computer.

Create a directory for your project, such as *console* :

```
mkdir console
cd console
```

Then add a file named *hello.js* with the following content:

```
#!/usr/bin/env node

// output message
console.log('Hello World!');
```

Save and run it from the command line:

```
node hello.js
```

You should now see *Hello World!* .

---

[1] https://nodejs.org/api/

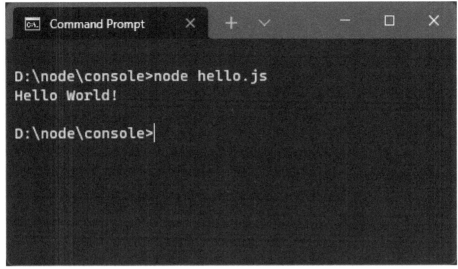

3-1. Hello World! in the console

#!/What?

The first line in `hello.js` — `#!/usr/bin/env node` —is known as a **shebang** or **hashbang**. It's entirely optional and ignored when you run `node hello.js`, because you're passing the script to the Node.js runtime which executes the code.

However, the shebang can be useful in Linux and macOS because it specifies which runtime to use—in this case, `node`. You can run the script using `./hello.js` alone but, before you can do that, you must permit direct execution by entering the following OS command in your terminal:

```
chmod +x ./hello.js
```

From then on you can run the script from the command line using this:

```
./hello.js
```

The OS analyses the shebang and runs the code using Node.js. It's not necessary to enter the full `node hello.js` command, although that will continue to work.

This is beyond the scope of Node.js and we won't use it again, because npm provides some cross-platform options. It's there should you need it.

To make the script more useful, you could pass a name on the command line. The process.argv[2] property in the standard library returns an array containing the command-line arguments:

- the first (element 0) is the `node` command itself
- the second (element 1) is the script you're running (`hello.js`)

[2.] https://nodejs.org/api/process.html

░ the third (element 2) is the first argument passed

Edit your *hello.js* script to extract the second argument and output it in the *console.log()* statement:

```
#!/usr/bin/env node

// fetch name from command or fallback
const nameArg = (process.argv[2] || 'world');

// output message
console.log(`Hello ${ nameArg }!`);
```

Save this, then run *node hello.js Craig* to see *Hello Craig!* .

3-2. Hello Craig! in the console

If you omit the parameter (*node hello.js*), the app falls back to *Hello world* .

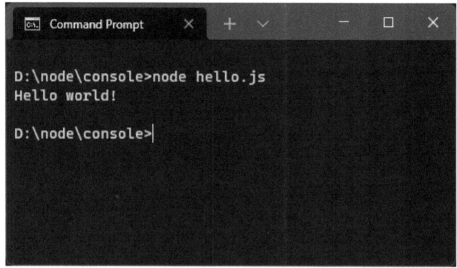

3-3. Hello world! is shown in the console

The fallback text of "world" is a little boring, so you could fetch the user's name stored in the OS's environment variables. The process.env[3] property returns an object containing environment variable name/value pairs. Try entering `process.env` in the REPL. (See the section "Run JavaScript Commands in the Node.js REPL" in Chapter 2 for more on this.)

Linux and macOS devices define a `USER` variable, while Windows sets `USERNAME`. Ensure your script is cross-platform by examining both when declaring `nameArg`:

```
// fetch name from command argument, environment, or fallback
const nameArg = (process.argv[2] || process.env.USER || process.env.USERNAME ||
↪ 'world');
```

Run the script with `node hello.js` and you'll see `Hello <yourname>`. You can still override your OS name by passing a parameter such as `node hello.js Craig`.

You can add a finishing touch to your console app by capitalizing the initial

---

[3] https://nodejs.org/api/process.html#process_process_env

letter of any name. Here's the final script:

```
#!/usr/bin/env node

// fetch name from command argument, environment, or fallback
const nameArg = capitalize(process.argv[2] || process.env.USER ||
↪process.env.USERNAME || 'world');

// output message
console.log(`Hello ${ nameArg }!`);

// capitalize the first letter of all words
function capitalize(str) {

  return str
    .trim()
    .toLowerCase()
    .split(' ')
    .map(word => word.charAt(0).toUpperCase() + word.slice(1))
    .join(' ');

}
```

Run the script with this:

```
node hello.js "from my node.js script"
```

You'll now see *Hello From My Node.js Script!* .

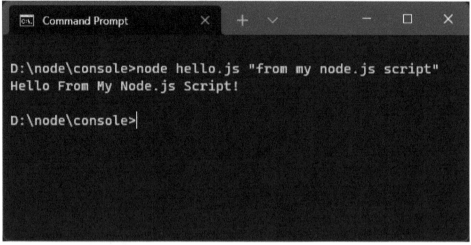

3-4. The hello response in the console

You can see a video demonstration of this in action here[4].

Your First Web Server App

Web applications require a web server to return HTML web pages when they're requested by a browser. The browser may then request other assets such as CSS stylesheets, client-side JavaScript, images, and Ajax-powered APIs.

Dedicated web servers such as Apache and NGINX are often used for this task. If Apache receives a request for a PHP file, it passes the content to the PHP interpreter, which runs the code. Apache receives the resulting output and returns it to the user's browser. PHP developers often access Apache server variables or tweak permissions to enhance their code.

---

[4]. https://spnt.co/nodevid3

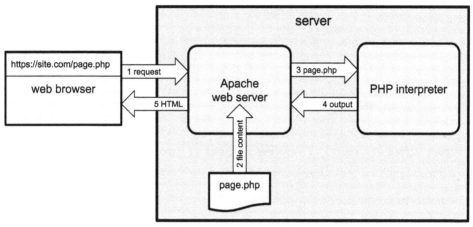

3-5. Apache/PHP web server

Node.js takes a different approach: *your JavaScript application is a web server.*
This sounds as though it's complex to code, but the HTTP[5] and HTTPS[6]
standard libraries do much of the work for you.

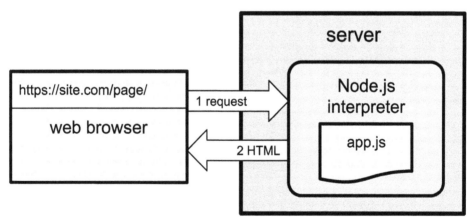

3-6. Node.js web server

Create a directory for your project, such as *server* :

```
mkdir server
```

---

[5]. https://nodejs.org/api/http.html

[6]. https://nodejs.org/api/https.html

```
cd server
```

Then add a file named *webhello.js* with the following content:

```
#!/usr/bin/env node

const
  port = (process.argv[2] || process.env.PORT || 3000),
  http = require('http');

http.createServer((req, res) => {

  console.log(req.url);

  res.statusCode = 200;
  res.setHeader('Content-Type', 'text/html');
  res.end(`<p>Hello World!</p>`);

}).listen(port);

console.log(`Server running at http://localhost:${ port }/`);
```

Run it with *node webhello.js* and you'll see *Server running at http://localhost:3000/* or similar. Open that address in your web browser to view a web page with a "Hello World!" paragraph.

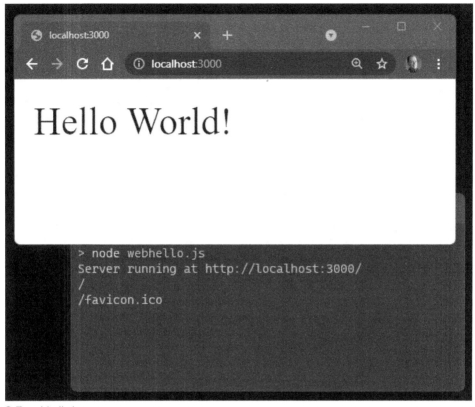

3-7. webhello.js

The code does the following:

- It defines a variable for the server's `port`. This can be passed on the command line, a `PORT` environment variable, or it falls back to `3000`.
- It uses the HTTP createServer[7] library to create a web server which listens on that `port`. When its callback function receives a request, it can examine the details in the `req` object and return a response using the `res` object.

This is a simple example, and the server returns the same "Hello World!" response regardless of the URL. Try accessing `http://localhost:3000/`, `http://localhost:3000/abc/`, or `http://localhost:3000/abc/123/`: every

---

[7] https://nodejs.org/api/http.html#http_http_createserver_options_requestlistener

page is the same.

 Port 3000?

> Web servers listen for HTTP requests on port 80 and HTTPS
> requests on port 443. You can set a different port, but you must
> specify it on the URL.
>
> Using the standard ports has drawbacks:
>
> - They may be in use by other software, such as other web servers or
>   Skype.
> - Linux and macOS block apps listening on ports below 1000 unless
>   they're launched by a superuser. This grants your script unlimited
>   rights, where it could do anything such as wiping your OS or posting
>   passwords to Twitter. Remember, you're running your code as well as
>   hidden code inside Node.js and any modules you've installed.
>
> It's safer to run web applications with standard permissions on a
> higher port. Live production servers can use a web server such as
> NGINX to forward requests to Node.js.

Let's improve the application by saying "hello" to a string passed on the URL.
The URL is available in `req.url`, so you can strip any non-word characters and
capitalize as before. Update the script to this:

```
#!/usr/bin/env node

const
  port = (process.argv[2] || process.env.PORT || 3000),
  http = require('http');

http.createServer((req, res) => {

  console.log(req.url);
  const nameArg = capitalize( req.url.replace(/[^\w.,-]/g, ' ').replace
  ↪(/\s+/g, ' ').trim() || 'world' );
```

```
   res.statusCode = 200;
   res.setHeader('Content-Type', 'text/html');
   res.end(`<p>Hello ${ nameArg }!</p>`);

}).listen(port);

console.log(`Server running at http://localhost:${ port }/`);

// capitalize the first letter of all words
function capitalize(str) {

  return str
    .trim()
    .toLowerCase()
    .split(' ')
    .map(word => word.charAt(0).toUpperCase() + word.slice(1))
    .join(' ');

}
```

Now open *http://localhost:3000/from/Node.js* in your browser. Chances are that you'll see "Hello World!", because your previous application instance is still running!

Switch to the terminal and press Ctrl | Cmd + C to stop the application. Restart it with *node webhello.js*, return to your browser, and refresh the page to see "Hello From Node.js!"

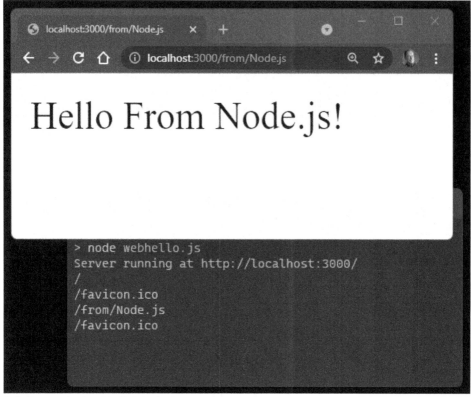

3-8. The result of webhello.js in the browser

Experiment with different URL paths and analyze the *nameArg* declaration to understand how it works.

Switch back to the Node.js terminal after you've tried a few URLs. It's logging each request, and you'll see something like this:

```
$ node webhello.js
Server running at http://localhost:3000/
/from/Node.js
/favicon.ico
/craig
/favicon.ico
```

What are those unexpected */favicon.ico* requests? You'll investigate further and debug in the next chapter.

Restarting Node.js Applications with Nodemon

You must restart a running Node.js application every time you make a change. Pressing `Ctrl` | `Cmd` + `C` and launching again will quickly become tiresome.

Nodemon[8] is a utility that monitors your source files for changes and automatically restarts the application. Install it globally with npm:

```
npm install -g nodemon
```

You can now use *nodemon* in place of *node* to launch any Node.js application. For example:

```
nodemon webhello.js
```

(You can pass any arguments as before.)

When you save a code change, Nodemon restarts the application and you'll see a log entry in the terminal:

```
[nodemon] restarting due to changes...
[nodemon] starting `node webhello.js`
```

If it doesn't work, try running *nodemon* with the *--legacy-watch* / *-L* argument:

```
nodemon -L webhello.js
```

Refer to the Nodemon documentation[9] for more options.

---

8. https://nodemon.io/
9. https://github.com/remy/nodemon

Executing Scripts from Windows Powershell

By default, Windows Powershell won't let you execute third-party scripts such as *nodemon* . Enter this command in a Powershell window to permit script execution:

```
Set-ExecutionPolicy RemoteSigned -Scope CurrentUser
```

You can find a video demonstration of the web application in action here[10].

Web Application Considerations

 Complexity Ahead

This section covers some advanced topics. Don't worry if it doesn't make sense now. We'll revisit the information later.

It's astonishing that this lightweight script implements a functional web server. The app is permanently on, and it can retain its own state regardless of the number of users. For example, it could establish a database connection once at start-up, then reuse that same connection on every request. Apache/PHP environments are usually stateless, so every page request must load configuration parameters and connect to a database before running a query.

However, Node.js applications run on a single processing thread:

- If your app fails, it fails for everyone and won't restart unless you have appropriate monitoring in place. Options including PM2[11] and forever[12] can help.
- If a single user triggers a long-running JavaScript function that takes ten

---

10. https://spnt.co/nodevid4
11. https://pm2.keymetrics.io/
12. https://www.npmjs.com/package/forever

seconds to complete, *every* user accessing at that time will be waiting at least ten seconds for a response. Asynchronous code solves the problem, but it takes time to understand the concepts.

Scaling an application can be difficult. Throwing more RAM or CPUs onto an Apache/PHP server will improve response times. Node.js still runs on a single CPU core even when that CPU has 15 more at its disposal. Solutions such as clustering[13], PM2, and Docker containers can help by launching multiple instances of the same application.

In addition, Node.js web servers are not efficient at serving static files such as images, stylesheets, and client-side JavaScript. Production sites often use a front-end NGINX web server to serve static files or direct the request to the Node.js application when appropriate. This is known as a *reverse proxy* and it has benefits, such as:

Static assets are served without any Node.js interaction. This avoids unnecessary processing and improves performance.

Settings such as HTTPS certificates can be handled by NGINX rather than Node.js. This is especially practical when you have more than one instance of the same Node.js application running.

A Node.js app can run on a port above 1000, so it doesn't need elevated superuser permissions.

---

[13] https://nodejs.org/api/cluster.html

Write Stateless Applications

Suppose your single Node.js app kept count of the number of logged-in users in single global variable named *userCount* .

What would happen if you wanted to improve performance by launching two or more instances of the same app—perhaps on other servers? Any instance could handle a user login. The *userCount* value would be different—*and wrong*—on each running instance of the app.

During development, you'll often work on a single running instance. However, I recommend you make it stateless to ensure it can scale and be more resilient. Always presume:

- multiple instances could be running anywhere, possibly on different ports or servers
- an instance can be started or stopped at any time
- a frontend web server will direct a single user's request to any instance—regardless of which instance handled a previous request

In essence, avoid storing application or user state in variables or local files that could differ across instances. Use a database to retain state so every instance of the application can be synchronized.

Summary

In this chapter, you've learned how to write simple console and web server applications using Node.js libraries alone. You've also seen how *nodemon* can automatically restart your app after updating code.

In the next chapter, you'll discover options for debugging and fixing problems in your Node.js code.

Quiz

1. Which of the following statements is true:

    a. Node.js can only run web apps.
    b. Node.js web apps require web server software such as NGINX to run.
    c. Node.js web apps don't require web server software, but NGINX or similar can be beneficial on production sites.
    d. Node.js isn't suitable for running production web applications.

2. Which steps are necessary after modifying a Node.js app?

    a. Use a tool such as *nodemon* to monitor for changes and restart the application.
    b. If it's already running, stop the application with `Ctrl` | `Cmd` + `C` and restart it.
    c. Close the terminal, open a new one, and start the application again.
    d. Any of the above.

3. Which Node.js object property returns command line arguments?

    a. `process.arg`
    b. `process.argv`
    c. `process.argument`
    d. `process.env`

4. Which Node.js object property returns environment variables?

    a. `process.env`
    b. `process.envv`
    c. `process.environment`
    d. `process.arg`

5. Can you launch multiple instances of the same Node.js app to improve resilience and performance?

- No. Only a single instance of a Node.js app can be launched at a time.
- Yes, but each instance must be on a separate real or virtual server.
- Yes, but containerization software such as Docker is essential.
- Yes, but the application should be stateless and receive requests via a load balancer or web server.

How to Debug Node.js Scripts

Chapter

4

Tutorials often describe debugging in the final chapters. This can be frustrating if you encounter a problem at the start of your coding journey—which you will. *Software development is complex.*

If you're lucky, your code will crash with an obvious error message. If you're unlucky, your application will carry on regardless but not generate the results you expect. If you're really unlucky, everything will work fine until the first user to arrive discovers a catastrophic, disk-wiping bug.

 Skip Ahead?

> This is a long chapter that describes several debugging options. You can skip ahead to the "Exercise: Debugging webhello.js" section (near the end of the chapter) if you'd like to get going. That said, a little learning now could save hours of effort later!

What is Debugging?

Debugging is the black art of fixing software defects. Fixing a bug is often easy; a corrected character or additional line of code solves the problem. *Finding* that bug is another matter, and developers can spend many frustrating hours locating the source of an issue. Fortunately, Node.js has some great tools to help trace problems.

How to Avoid Bugs

You can often prevent bugs before you test your application. Let's look at some ways.

Use a Good Code Editor

A good code editor offers features such as:

- line numbering to locate where an error occurred
- type checking—for example, to ensure a number variable can't have a

string assigned

- color-coding to catch syntax issues, such as invalid statements or missing quotes
- auto-completion of variable names, function names, properties, etc.
- bracket matching to highlight problems in nested structures
- auto-indentation that uses the correct tab or space characters
- variable renaming across files and projects
- snippet saving and reuse
- parameter prompts for functions, properties, and methods
- function navigation to jump to a declaration
- unreachable code detection
- refactoring tools to rearrange messy code

Node.js developers are spoiled for choice, with editors such as VS Code[1], Atom[2], and Sublime Text[3].

Use a Code Linter

A **linter** reports faults such as syntax errors, poor indentation, undeclared variables, mismatching brackets, and your own preferences (semicolons, quote usage, etc.) before you save and test your code. Popular options for JavaScript and Node.js include ESLint[4], JSLint[5], and SHint[6]

These can be installed as global Node.js modules. For example, here's how to install ESLint globally using npm:

```
npm install eslint -g
```

You can then check files from the command line:

---

[1]. https://code.visualstudio.com/
[2]. https://atom.io/
[3]. https://www.sublimetext.com/
[4]. https://eslint.org/
[5]. https://www.jslint.com/
[6]. https://jshint.com/

```
eslint myscript.js
```

```
craig@craigdev: ~                    ×    +  ∨                        —  □  ×
craig@craigdev:~$ more myscript.js
const counter = 10;

for (i = 0; i < countr; i++) {
  console.log(i)
}
craig@craigdev:~$ eslint myscript.js

/home/craig/myscript.js
   1:7    error   'counter' is assigned a value but never used   no-unused-vars
   3:6    error   'i' is not defined                             no-undef
   3:13   error   'i' is not defined                             no-undef
   3:17   error   'countr' is not defined                        no-undef
   3:25   error   'i' is not defined                             no-undef
   4:15   error   'i' is not defined                             no-undef
   4:17   error   Missing semicolon                              semi

✗ 7 problems (7 errors, 0 warnings)
  1 error and 0 warnings potentially fixable with the `--fix` option.

craig@craigdev:~$ |
```

4-1. Using ESLint from the command line

However, most linters have code editor plugins, such as ESLint for VS Code[7] and linter-eslint for Atom[8], which check your code as you type.

[7] https://marketplace.visualstudio.com/items?itemName=dbaeumer.vscode-eslint
[8] https://atom.io/packages/linter-eslint

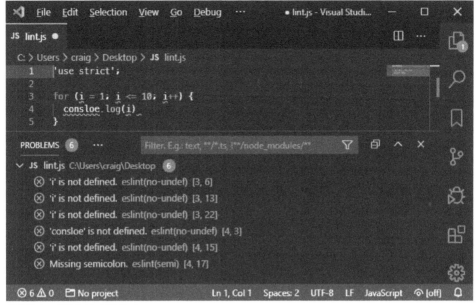

4-2. Using ESLint in VS Code

Use Source Control

A source control system such as Git[9] can help safeguard your code and manage revisions. It becomes easier to discover where and when a bug was introduced and who should receive the blame! Online repositories such as GitHub[10] and Bitbucket[11] offer free space and management tools.

Adopt an Issue-tracking System

Does a bug exist if no one knows about it? An issue-tracking system is used to report bugs, find duplicates, document reproduction steps, determine severity, calculate priorities, assign developers, record discussions, and track progress of fixes.

Online source code repositories often offer basic issue tracking, but dedicated

---

[9.] https://git-scm.com/
[10.] https://github.com/
[11.] https://bitbucket.org/

solutions such as Jira[12], FogBugz[13], or Bugzilla[14] may be appropriate for larger teams and projects.

Use Test-driven Development

Test-driven development (TDD) is a development process that encourages developers to write code to test the operation of a function *before* that function is written—as in *is X returned when function Y is passed input Z?*

Tests are run as you develop code to prove the resulting function works as expected. The same test can be rerun to spot issues as further changes are made. *Of course, your tests could have bugs too!*

Further resources:

- TDD overview at Wikipedia[15]
- "What is Test Driven Development"[16]
- "Master Test-driven Development in Node.js"[17]

Node.js Debugging Environment Variables

Environment variables[18] set within the host operating system control Node.js application settings. The most common is `NODE_ENV`, which is typically set to `development` when debugging or `production` on a live server.

Environment variables can be set on Linux/macOS:

[12.] https://www.atlassian.com/software/jira

[13.] https://fogbugz.com

[14.] https://www.bugzilla.org

[15.] https://en.wikipedia.org/wiki/Test-driven_development

[16.] https://www.browserstack.com/guide/what-is-test-driven-development

[17.] https://www.sitepoint.com/premium/courses/master-test-driven-development-in-node-js-2932

[18.] https://nodejs.org/api/cli.html#environment-variables

```
NODE_ENV=development
```

This is the Windows Command Prompt:

```
set NODE_ENV=development
```

This is for Windows Powershell:

```
$env:NODE_ENV="development"
```

Internally, your own application can detect the setting and enable debugging messages when necessary. For example:

```
// running in development mode?
const DEVMODE = (process.env.NODE_ENV !== 'production');

if (DEVMODE) {
  console.log('application started in development mode');
}
```

NODE_DEBUG enables debugging messages using the Node.js *util.debuglog* . (See the "Node.js *util.debuglog* " section below.) You should also consult the documentation of your primary modules and frameworks to discover further logging options.

Node.js Debugging Command-line Options

Various command-line options[19] can be passed to the *node* or *nodemon* runtime when launching an application. One of the most useful is *--trace-warnings* , which outputs stack traces when promises don't resolve or reject as expected:

```
node --trace-warnings index.js
```

[19.] https://nodejs.org/api/cli.html

Other options include:

- `--enable-source-maps` : enable source maps when using a transpiler such as TypeScript
- `--throw-deprecation` : throw errors when deprecated features are used
- `--inspect` : activate the V8 inspector (see the "Node.js V8 Inspector" section below)

Console Debugging

One of the easiest ways to debug an application is to output values to the console during execution:

```
console.log( myVariable );
```

 Never Use console.log()?!

Some developers claim you should *never* use `console.log()` , because you're changing code and there are better debugging options. This is true—*but everyone does it!*

Use whatever tool makes you productive. Console logging can be a quick and practical option. Finding a bug is more important than the method you used to find it.

Few developers delve beyond the standard `console.log()` command, but they're missing out on many more possibilities[20]:

[20.] https://nodejs.org/api/console.html

`console` method	Description
`.log(msg)`	output a message to the console
`.log('%j', obj)`	output an object as a compact JSON string
`.dir(obj,opt)`	uses `util.inspect` to pretty-print objects and properties
`.table(obj)`	outputs arrays of objects in tabular format
`.error(msg)`	output an error message
`.count(label)`	a named counter reporting the number of times the line has been executed
`.countReset[label]`	resets a named counter
`.group(label)`	indents a group of log messages
`.groupEnd(label)`	ends the indented group
`.time(label)`	starts a timer to calculate the duration of an operation
`.timeLog([label]`	reports the elapsed time since the timer started
`.timeEnd(label)`	stops the timer and reports the total duration
`.trace()`	outputs a stack trace (a list of all calling functions)
`.clear()`	clear the console

console.log() accepts a list of comma-separated values. For example:

```
let x = 123;
console.log('x:', x);
// x: 123
```

However, ES6 destructuring[21] offers similar output with less typing effort:

```
console.log({ x });
// { x: 123 }
```

[21.] https://www.sitepoint.com/es6-destructuring-assignment/

util.inspect can format objects for easier reading, but *console.dir()* does the hard work for you:

```
console.dir(obj, { depth: null, color: true });
```

Node.js util.debuglog

The Node.js *util* module offers a built-in *debuglog* [22] method that conditionally writes log messages to *STDERR* :

```
const util = require('util');
const debuglog = util.debuglog('myapp');

debuglog('myapp debug message [%d]', 123);
```

When the *NODE_DEBUG* environment variable is set to *myapp* (or a wildcard such as *** or *my**), debugging messages are displayed in the console:

```
MYAPP 9876: myapp debug message [123]
```

(*9876* is the Node.js process ID.)

Debugging with Logging Modules

Third-party logging modules are available should you require more sophisticated options for messaging levels, verbosity, sorting, file output, profiling, reporting, and more. Popular solutions include:

- cabin[23]
- loglevel[24]
- morgan[25] (Express middleware)

---

[22] https://nodejs.org/api/util.html#utildebuglogsection-callback
[23] https://www.npmjs.com/package/cabin
[24] https://www.npmjs.com/package/loglevel
[25] https://www.npmjs.com/package/morgan

- pino[26]
- signale[27]
- storyboard[28]
- tracer[29]
- winston[30]

Node.js V8 Inspector

The following sections use the `webhello.js` script developed in the previous chapter to illustrate debugging concepts.

Node.js is a wrapper around the V8 JavaScript engine. V8 includes its own inspector and debugging client[31]. Use the `inspect` argument to start an application (not to be confused with the `--inspect` flag—which is covered below in the "Debugging Node.js Apps with Chrome" section):

```
node inspect webhello.js
```

The debugger pauses at the first line and displays a `debug` prompt:

```
$ node inspect webhello.js
< Debugger listening on ws://127.0.0.1:9229/8bf7669c-b3b4-43e6-9f96-3b40abbcb479
< For help, see: https://nodejs.org/en/docs/inspector
<
connecting to 127.0.0.1:9229 ... ok
< Debugger attached.
<
Break on start in webhello.js:4
  2
  3 const
> 4   port = (process.argv[2] || process.env.PORT || 3000),
```

---

26. https://www.npmjs.com/package/pino
27. https://www.npmjs.com/package/signale
28. https://www.npmjs.com/package/storyboard
29. https://www.npmjs.com/package/tracer
30. https://www.npmjs.com/package/winston
31. https://nodejs.org/api/debugger.html

```
5    http = require('http');
6
```

Enter `help` to view a list of commands. You can step through the application with these options:

- `cont` or `c` : continue execution
- `next` or `n` : run the next command
- `step` or `s` : step into a function being called
- `out` or `o` : step out of a function and return to the calling command
- `pause` : pause running code

Other options include:

- watching variable values with `watch('myvar')` [32]
- setting breakpoints with the `setBreakpoint()` / `sb()` command[33] (although it's easier to insert a `debugger;` statement in your code)
- `restart` a script
- `.exit` the debugger (the initial `.` is required)

If this sounds horribly clunky, *it is*. Only use the built-in debugging client when there's absolutely no other option or you're feeling masochistic.

Debugging Node.js Apps with Chrome

Start the Node.js V8 inspector with the `--inspect` flag:

```
node --inspect webhello.js
```

(`nodemon` can be run instead of `node` if necessary.)

This starts the debugger listening on `127.0.0.1:9229` , which any local debugging client can attach to:

---

[32] https://nodejs.org/api/debugger.html#debugger_watchers
[33] https://nodejs.org/api/debugger.html#debugger_breakpoints

```
Debugger listening on ws://127.0.0.1:9229/20ac75ae-90c5-4db6-af6b-d9d74592572f
```

If you're running the Node.js application on another device or Docker container, ensure port *9229* is accessible and grant remote access using this:

```
node --inspect=0.0.0.0:9229 webhello.js
```

Alternatively, use *--inspect-brk* to halt processing the first statement so you can step through the application line by line.

Open the Chrome browser and enter *chrome://inspect* in the address bar.

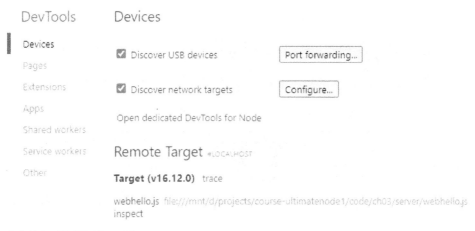

4-3. Using Chrome inspect

 **Not Using Chrome?**

> Chromium, Edge, Opera, Vivaldi, and Brave all have the same debugger as Chrome. The *chrome://inspect* address should work identically.

Remote Target

If the Node.js application doesn't appear as a **Remote Target**, ensure **Discover network targets** is checked, then click **Configure** to add the IP address and port of the device where the application is running.

Click the Target's **inspect** link to launch DevTools. This will be immediately familiar to anyone who's used browser developer tools.

In the **Sources** pane, click **+ Add folder to workspace**, select where your Node.js files are located, and hit **Agree**. Open `webhello.js` in the left-hand pane or by pressing Ctrl | Cmd + P.

4-4. Chrome DevTools

Click any line number to set a breakpoint denoted by a blue marker. A **breakpoint** specifies where the debugger should pause processing so you can inspect the state of the program. You can define any number of breakpoints.

Debugger Statement

Processing also halts at any *debugger* statement in your code when it runs using the V8 inspector. This may be practical when sharing code or debugging across multiple devices, although you may want to remove those commands before committing the code to source control or releasing on a live server.

Refresh/open http://localhost:3000/ in your browser and code execution stops when that breakpoint is reached.

4-5. Chrome DevTools breakpoint

The right-hand panels include:

- a **Watch** pane, which allows you to monitor variables by clicking the + icon and entering their name
- a **Breakpoints** pane, which shows a list of all breakpoints and allows them to be enabled or disabled
- a **Scope** pane, which shows the state of all available local and global

variables

- a **Call Stack** pane, which shows the functions that were called to reach this point

The row of icons above the **Paused on breakpoint** message is pictured below.

4-6. Chrome breakpoint icons

These options perform the following actions (from left to right):

1   **resume execution**: continue processing until the next breakpoint

2   **step over**: execute the next command but stay within the current

function; don't jump into any function it calls

3   **step into**: execute the next command and jump into any function it calls

4   **step out**: continue processing to the end of the function and return to

the calling command

5   **step**: similar to **step into**, except it won't jump into *async* functions

6   **deactivate all breakpoints**

7   **pause on exceptions**: halt processing whenever an error occurs

You can find a video demonstration of debugging with Chrome here.[34]

Debugging Node.js Apps with VS Code

Node.js debugging in VS Code requires no configuration when you run a Node.js application on your local system. Open the starting file (use `webhello.js` here), activate the **Run and Debug** pane, click the **Run and**

---

[34] https://spnt.co/nodevid5

Debug Node.js button, and choose the Node.js environment.

4-7. The VS Code debugger

The debugging screen is similar to the DevTools screen, with a **Variables**, **Watch**, **Call stack**, **Loaded scripts**, and **Breakpoints** list. Set a breakpoint by clicking the left-hand gutter next to the line number. A red dot icon appears. Refresh http://localhost:3000/ in your browser and execution will halt on the breakpoint line so you can examine the program state.

4-8. A VS Code breakpoint

The icons in the debugging toolbar at the top are used to resume execution, step over, step into, step out, restart, or stop the application. Identical options are available from the **Run** menu.

You can also right-click a line number.

4-9. VS Code breakpoint options

Once you've done that, you can set the following:

- A standard breakpoint.
- A conditional breakpoint that halts the program when criteria are met—such as `count > 3`.
- A logpoint. This is effectively `console.log()` without code! Enter any string with evaluated expressions in curly braces. For example, `URL: { req.url }` outputs the value of the `req.url` property.

```
 9      // console.log(req.url);
10      const nameArg = capitalize( r
11
12      res.statusCode = 200;

Log Message ∨   URL: { req.url }|

13      res.setHeader('Content-Type',
14      res.end(`<p>Hello ${ nameArg
```

4-10. A VS Code logpoint

The DEBUG CONSOLE displays the logpoint value when the web page is refreshed.

```
TERMINAL    PROBLEMS    OUTPUT    DEBUG CONSOLE

URL: /
URL: /favicon.ico
```

4-11. VS Code logpoint output

For more information, refer to "Debugging in Visual Studio Code"[35].

Advanced Debugging Configuration

VS Code configuration is necessary when you're debugging code on another device, a virtual machine, or you want to use different launch options. VS Code stores launch configurations in a `launch.json` file inside a `.vscode` folder in

---

[35]. https://code.visualstudio.com/docs/introvideos/debugging

your project. To generate the file, click the **create a launch.json file** link at the top of the **Run and Debug** pane and choose the **Node.js** environment.

```
JS webhello.js  U        launch.json  U                                        ⠿ ⏸  ⤴  ⤵ ↑  ⟲ ⬚                            
.vscode > ⟩ launch.json > …
  1   {
  2       // Use IntelliSense to learn about possible attributes.
  3       // Hover to view descriptions of existing attributes.
  4       // For more information, visit: https://go.microsoft.com/fwlink/?linkid=830387
  5       "version": "0.2.0",
  6       "configurations": [
  7           {
  8               "console": "integratedTerminal",
  9               "internalConsoleOptions": "neverOpen",
 10               "name": "nodemon",
 11               "program": "${workspaceFolder}/webhello.js",
 12               "request": "launch",
 13               "restart": true,
 14               "runtimeExecutable": "nodemon",
 15               "skipFiles": [
 16                   "<node_internals>/**"
 17               ],
 18               "type": "pwa-node"
 19           }
 20       ]
 21   }
 22
                                                                                                        Add Configuration…
```

4-12. VS Code launch configuration

You can add any number of configuration setting objects to the `"configurations": []` array. Click the **Add Configuration** button to add an appropriate option. VS Code can either:

- **Launch** a process using Node.js itself
- **Attach** to a Node.js inspector process, perhaps running on a remote machine or Docker container

The screenshot above shows a `nodemon` launch configuration. The **Add Configuration** button provides a `nodemon` option; it's only necessary to edit the `"program"` property to point at `${workspaceFolder}/webhello.js`.

Save `launch.json`, then select `nodemon` from the drop-down list at the top of the **Run and Debug** pane, and click the green run icon.

4-13. VS Code launch

nodemon will launch your application. You can edit the code and set breakpoints or logpoints as before.

For further information, refer to the VS Code launch configurations[36].

VS Code can debug any Node.js application, but the following extensions can make life easier:

- Remote - Containers[37]: connect to apps running in Docker containers
- Remote - WSL[38]: connect to apps running on Linux in WSL on Windows

Other Node.js Debugging Tools

The Node.js Debugging Guide[39] provides advice for other IDEs and editors including Visual Studio, JetBrains, WebStorm, Gitpod, and Eclipse. Atom also has a node-debug[40] extension.

ndb[41] offers an *improved debugging experience* with powerful features such as attaching to child processes and script blackboxing so that only code in

---

[36]. https://code.visualstudio.com/docs/editor/debugging#_launch-configurations
[37]. https://marketplace.visualstudio.com/items?itemName=ms-vscode-remote.remote-containers
[38]. https://marketplace.visualstudio.com/items?itemName=ms-vscode-remote.remote-wsl
[39]. https://nodejs.org/en/docs/guides/debugging-getting-started/
[40]. https://atom.io/packages/node-debug
[41]. https://github.com/GoogleChromeLabs/ndb

specific folders is shown.

The IBM report-toolkit for Node.js[42]works by analyzing data output when `node` runs with an `--experimental-report` option.

Finally, commercial services such as LogRocket[43] and Sentry.io[44] integrate with your live web application on both the client and the server to record errors as they're encountered by real users.

Exercise: Debugging webhello.js

The `webhello.js` code has a strange bug where an unexpected `/favicon.ico` request is logged. To examine what's happening, launch VS Code and open the folder containing `webhello.js` . Then:

1   Switch to the **Run and Debug** pane.

2   Click **create a launch.json file** and choose the **Node.js** environment.

3   Click the **Add Configuration** button and choose **Node.js: Nodemon setup**. (You'll now have two objects inside the `"configurations"` array. You can delete the second one.)

4   Change the `"program"` value to `"${workspaceFolder}/webhello.js"` .

5   Save the file and open `webhello.js` .

6   Click the **+** icon in the **Watch** pane and add the expression `req.url` .

7   Click the **+** icon in the **Watch** pane and add the expression `nameArg` .

8   Add a breakpoint to the line starting `res.end` by clicking to the left of the line number. A red circle icon will appear.

---

[42]. https://github.com/ibm/report-toolkit
[43]. https://logrocket.com/
[44]. https://sentry.io/

9   Click the **nodemon** green start icon at the top of the **Run and Debug**

pane.

4-14. VS Code launch

10   The web application will start.

Now open http://localhost:3000/ in your browser and processing will halt at
the breakpoint. Assuming you haven't used a different path on the URL, the
Watch pane will show:

```
req.url: `/`
nameArg: `World`
```

Click the **Continue** icon or press F5 to resume processing. At this point, the
browser will show "Hello World!"—but the breakpoint will trigger again.

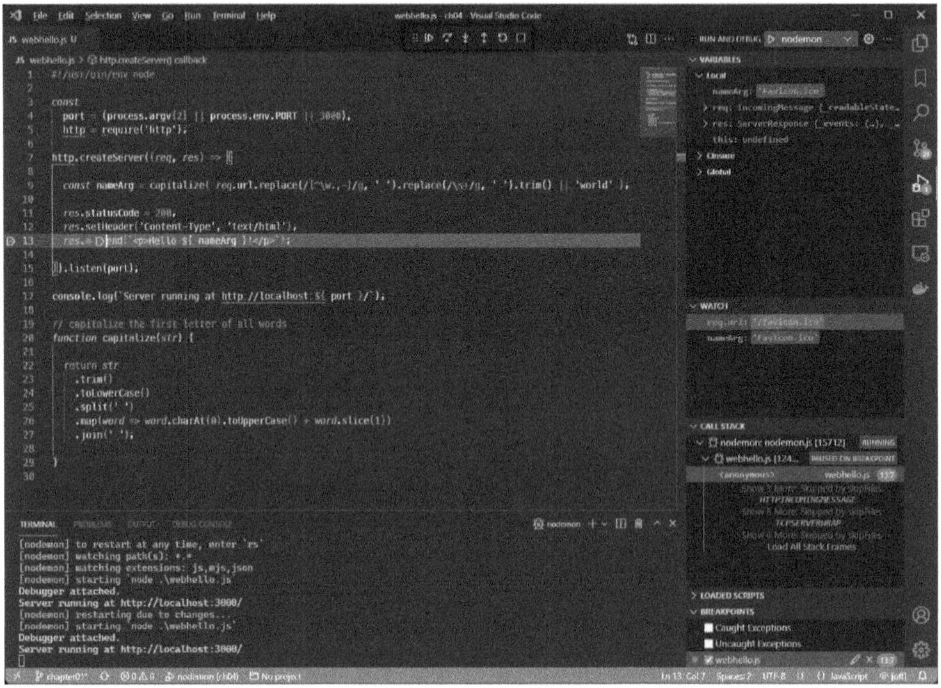

4-15. VS Code breakpoint view

The **Watch** pane shows:

```
req.url: `/favicon.ico`
nameArg: `Favicon.ico`
```

(If this doesn't happen, try a hard refresh in your browser—usually Ctrl + F5 on Windows and Linux or Cmd + R on macOS.)

When a browser makes its first request for a web page, it also requests a *favicon.ico* image. This is the icon shown to the left of the page's title in the browser tab.

A web server would normally send an appropriate image or return an HTTP *404 Not found* error. However, your Node.js application treats it like any other request and returns the HTML text *"Hello Favicon.ico"*, which the browser can't use.

It's not a catastrophic bug, but both the browser and server are doing unnecessary work. Fix it by adding a check at the top of the *createServer* callback function, which returns a 404 error:

```
http.createServer((req, res) => {

  // abort favicon.ico request
  if (req.url.includes('favicon.ico')) {
    res.statusCode = 404;
    res.end('Not found');
    return;
  }
```

Save *webhello.js* , and *nodemon* will restart the application. Try refreshing your browser again and the breakpoint triggers just once.

To finish debugging, click the red square **Stop** icon in the debugging toolbar.

You can find a video demonstration of debugging with VS Code here[45].

Summary

This chapter has illustrated options for debugging Node.js applications. Use whatever makes you productive, but I generally use console logging for quick and dirty bug hunting and VS Code when things get complicated.

In the next chapter, you'll start to write more complex Node.js code using npm and third-party modules.

Debugging Terminology

Debugging has its own selection of obscure jargon. We've covered most aspects throughout this chapter, but you could encounter terms like the ones shown below.

[45.] https://spnt.co/nodevid6

Term	Explanation
breakpoint	a line at which a debugger halts a program so its state can be inspected
breakpoint (conditional)	a breakpoint triggered by a certain condition, such as a value reaching 100. Also known as a **watchpoint**
debugger	a tool that offers debugging facilities such as running code line by line to inspect internal variable states
duplication	a reported bug that has already been reported—perhaps in a different way
feature	as in the claim: "it's not a bug, it's a feature". You'll find yourself saying this at some point
frequency	how often a bug occurs
it doesn't work	the most-often made but least useful bug report
logpoint	a debugger instruction that shows the value of an expression during execution
logging	output of runtime information to the console or a file
logic error	the program works but doesn't act as intended
priority	where a bug is allocated on a list of planned updates
race condition	hard-to-trace bugs dependent the sequence or timing of uncontrollable events
refactoring	rewriting code to help readability and maintenance
regression	re-emergence of a previously fixed bug perhaps owing to other updates
related	a bug that's similar or related to another
reproduce	the steps required to cause the error
RTFM error	user incompetence disguised as a bug report, typically followed by a developer's response that they should "Read

Term	Explanation
	The *Friendly* Manual"
step into	when running code line by line in a debugger, step into the function being called
step out	when running line by line, complete execution of the current function and return to the calling code
step over	when running line by line, complete execution of a command without stepping into a function it calls
severity	the impact of a bug on system. For example, data loss would normally be considered more problematic than a one-pixel UI alignment issue unless the frequency of occurrence is very low
stack trace	the historical list of all functions called before the error occurred
syntax error	typographical errors, such as `console.lug()`
user error	an error caused by a user rather than the application. This may still incur an update, depending on the seniority of the person who caused it!
watch	a variable or expression output during debugger execution

Quiz

1. You can debug Node.js apps by:

    a. using the command-line V8 inspector
    b. attaching to the process using Chrome DevTools
    c. using a suitable editor such as VS Code
    d. all of the above

2. What command would be suitable for outputting the values contained in a JavaScript object?

a. `console.log('%j', obj)`
b. `console.table(obj)`
c. `console.dir(obj, { depth: null, color: true })`
d. any of the above

3. A breakpoint is:

a. triggered by `console.log()`
b. a point at which processing halts during execution
c. a statement to stop the program, such as `exit`
d. the moment a developer chooses to stop work

4. A logpoint is:

a. used to show the value of an expression during execution
b. an alternative name for a breakpoint
c. a reference to a `console.log()` statement
d. a specific line in an output log

5. `console.log()` :

a. should never be used
b. should only be used when there's no other option
c. should be used if it'll help locate a bug
d. is impractical for debugging

Getting Started with Express

Chapter

5

In this chapter, you'll create a web server application that constructs and returns simple web pages. It will help you become more familiar with:

- npm (Node Package Manager)
- ES6 modules
- the Express[1] framework
- URL routing
- HTML template engines

Why use Express?

You created a small web server application in Chapter 3. It's fast, and it works well, but a complex web application requires features such as URL routing, query string parsing, posted data decoding, HTML templates, image serving, and more. You could write this yourself, but much of that effort is already implemented in Express.

Express is a Node.js web server framework that promotes itself as "fast, unopinionated, and minimalist". It allows you to concentrate on your application's business logic without having to worry too much about web server technicalities such as URL routing, parsing data, setting HTTP headers, and so on.

Various web server frameworks are available in the Node.js ecosystem, including Fastify[2], Koa[3], and Hapi[4]. These may be more recent, more regularly maintained, faster, and a better fit for your application. However, Express was one of the first web frameworks and influenced all that followed. It's stable, easy to use, and remains popular, with 18 million downloads per week. You're more likely to encounter Express than another framework.

[1.] https://expressjs.com/
[2.] https://www.fastify.io/
[3.] https://koajs.com/
[4.] https://hapi.dev/

Express Version

> At the time of writing, Express 4 is the active recommended release
> and Express 5 is in alpha. All the examples below *should* work in
> either, but switch to version 4 if you have problems.

Create a New Node.js Project

Create and access a project directory for your new application. A name such
as *express* is fine:

```
mkdir express
cd express
```

 Create a New Git Repository

> For real projects, I recommend creating a new Git repository and
> cloning it accordingly. This is easier than attempting to *Git-ify* a
> partially written project later.

Run `npm init` to initialize a new Node.js project. npm will prompt you for
values, but you can hit `Enter` to accept the defaults.

```
craig@craigdev:~/apps$ mkdir express
craig@craigdev:~/apps$ cd express/
craig@craigdev:~/apps/express$ npm init
This utility will walk you through creating a package.json file.
It only covers the most common items, and tries to guess sensible defaults.

See `npm help init` for definitive documentation on these fields
and exactly what they do.

Use `npm install <pkg>` afterwards to install a package and
save it as a dependency in the package.json file.

Press ^C at any time to quit.
package name: (express)
version: (1.0.0)
description: Example Express.js app
entry point: (index.js)
test command:
git repository:
keywords:
author: Craig Buckler
license: (ISC) MIT
About to write to /home/craig/apps/express/package.json:

{
  "name": "express",
  "version": "1.0.0",
  "description": "Example Express.js app",
  "main": "index.js",
  "scripts": {
    "test": "echo \"Error: no test specified\" && exit 1"
  },
  "author": "Craig Buckler",
  "license": "MIT"
}


Is this OK? (yes) yes
craig@craigdev:~/apps/express$ |
```

5-1. Initializing a new Node.js project

npm saves the settings to a new *package.json* file in your project's root directory:

```
{
  "name": "express",
```

```
  "version": "1.0.0",
  "description": "Example Express app",
  "main": "index.js",
  "scripts": {
    "test": "echo \"Error: no test specified\" && exit 1"
  },
  "author": "Craig Buckler",
  "license": "MIT"
}
```

package.json provides a single place to configure your application. It contains the *name* , the *version* , the *main* entry/starting script, useful application *scripts* , configuration data, and module dependencies.

 Semantic Versioning

Most Node.js projects use semantic versioning[5], with three *MAJOR.MINOR.PATCH* numbers such as *1.2.33* . When a change occurs, you increment the appropriate number and zero those that follow:

- *MAJOR* for major updates with incompatible API changes
- *MINOR* for new functionality that doesn't affect backwards compatibility
- *PATCH* for bug fixes

Switch to ES6 Modules

Ensure your project uses standard ES6 modules by adding *"type": "module",* to *package.json* in your editor (it can go anywhere in the root object, but is placed above *"main"* here):

```
{
  "name": "express",
```

```
  "version": "1.0.0",
  "description": "Example Express app",
  "type": "module",
  "main": "index.js",
  "scripts": {
    "test": "echo \"Error: no test specified\" && exit 1"
  },
  "author": "Craig Buckler",
  "license": "MIT"
}
```

ES6 modules are identical to those used in web browsers. Node.js uses CommonJS by default, but ES6 support arrived in version 13. ES6 modules will become predominant over time, so we'll use ES6 throughout this course.

Node can import CommonJS modules using ES6 syntax. It will also make suggestions if there's a potential issue or conflict. However, you may encounter problems with some modules written in CommonJS syntax, especially if they haven't been updated for a few years.

Install Express

Install Express from your project directory using npm:

```
npm install express
```

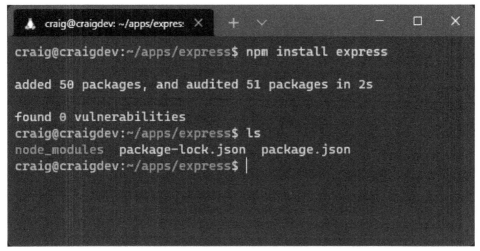

5-2. Installing Express

After completion, your *package.json* file will have a new *"dependencies"* object, which lists the modules required when your project runs. It contains a reference to *"express"* and its latest version number (the leading ^ means Express can upgrade to a compatible version such as *4.17.2* or *4.18.0* but not *5.0.0*):

```
{
  "name": "express",
  "version": "1.0.0",
  "description": "Example Express app",
  "type": "module",
  "main": "index.js",
  "scripts": {
    "test": "echo \"Error: no test specified\" && exit 1"
  },
  "author": "Craig Buckler",
  "license": "MIT",
  "dependencies": {
    "express": "^4.17.1"
  }
}
```

You'll also find the following:

- a new `package-lock.json` file for npm internal use, which lists all the installed modules
- a new `node_modules` folder, which contains the Express module and all submodules code (around 2MB of files)

 Runtime Dependencies and Development Dependencies

A module such as Express is required for your application to run. It's a **dependency**.

You can also install **development dependencies**, which are typically build tools that are only required on your development PC. Examples include JavaScript bundlers such as Rollup[6], CSS preprocessors such as Sass[7], and live reload systems such as Browsersync[8].

npm presumes a module is a runtime dependency unless you add the `--save-dev` switch during installation. For example:

```
npm install browser-sync --save-dev
```

This installs Browsersync, but references it in a `"devDependencies"` object in `package.json`. Running `npm install` on a production server where the `NODE_ENV` environment variable is set to `production` would *not* install Browsersync.

The distinction between a dependency and a development dependency is not always straightforward. For example, you could run Rollup on a production server to create minified JavaScript files.

Create the Express Entry Script

You can now write code that uses Express to create a web application. Add a

---

6. https://rollupjs.org/
7. https://sass-lang.com/
8. https://browsersync.io/

new *index.js* file in the project directory with the following code:

```js
// Express application
import express from 'express';

// configuration
const
  cfg = {
    port: process.env.PORT || 3000
  };

// Express initiation
const app = express();

// home page route
app.get('/', (req, res) => {
  res.send('Hello World!');
});

// start server
app.listen(cfg.port, () => {
  console.log(`Example app listening at http://localhost:${ cfg.port }`);
});
```

To make starting this app a little easier, edit *package.json* and change the
"scripts" object to this:

```json
"scripts": {
  "start": "nodemon index.js"
},
```

If you don't have *nodemon* installed, you can install it globally with *npm install nodemon -g* . If you'd rather use *node* directly, use *"node index.js"* as your *"start"* script (but you'll need to stop and restart your app every time you want to test a change).

Start the application with *npm start* and browse to <u>http://localhost:3000</u>.

5-3. Express start

The script imports the *express* module and creates an instance named *app*.

A single routing function is defined to handle HTTP GET requests to the root `/` path:

```
// home page route
app.get('/', (req, res) => {
```

```
  res.send('Hello World!');
});
```

 ## What Is Routing?

Routing determines which functions Express executes when it receives a request for a specific URL, such as / or /another/ path/ .

Ultimately, one function will return an HTTP response and terminate further processing. The order of your routing functions is therefore critical: a function won't run if an earlier function completes the request.

A routing function is passed these two objects:

- An Express HTTP Request object[9] (*req*), which contains details about the browser's request.
- An Express HTTP Response object[10] (*res*), which provides methods used to return a response to the browser. It sends "Hello, World!" text.

Try adding another routing function below the / handler to handle HTTP GET requests to /hello/ :

```
// another route
app.get('/hello/', (req, res) => {
  res.send('Hello again!');
});
```

Once the application has restarted, open http://localhost:3000/hello/ in your browser to see a "Hello again!" message.

No other URL routes are defined. Entering a different URL path in the

---

9. https://expressjs.com/en/4x/api.html#req
10. https://expressjs.com/en/4x/api.html#res

browser—such as http://localhost:3000/abc—returns *Cannot GET /abc* . Routing is a central part of Express[11], and the framework provides options for parsing and responding to different URLs.

The end of the script has an `app.listen()` call to start the Express server listening on the defined port.

See the course `code/ch05/express01` directory[12] and associated video[13] to run the code created so far.

Should You Switch to HTTPS?

Probably not.

All the Node.js examples in this course respond to HTTP requests on port `3000` by default:

```
// start server
app.listen(cfg.port, () => {
  console.log(`Example app listening at http://localhost:${ cfg.port }`);
});
```

HTTPS requires a Secure Socket Layer (SSL) certificate. These are issued by certificate authorities for use on a specific domain to encrypt tamper-proof data between the browser and server.

For local testing, developers often create their own self-signed certificates using the command line[14] or online tools[15].

If you have a private key file named *server.key* , and a site certificate named *server.crt* , an Express app can read the SSL files[16], create an HTTPS

---

11. https://expressjs.com/en/4x/api.html#router
12. https://github.com/spbooks/ultimatenode1/tree/main/ch05/express01
13. https://spnt.co/nodevid7
14. https://linuxize.com/post/creating-a-self-signed-ssl-certificate/
15. https://www.selfsignedcertificate.com/

server[17], and pass the Express *app* object as a listener:

```
// start HTTPS server
import { createServer } from 'https';
import { readFileSync } from 'fs';

createServer(
  {
    key: fs.readFileSync('./server.key'),
    cert: fs.readFileSync('./server.crt')
  },
  app
).listen(cfg.port);
```

(This replaces the HTTP `app.listen()` code above.)

Your application will now accept requests to
https://localhost:3000/—*although your browser will warn that the certificate has not been issued by a recognized Authority.*

Problems with this approach include the following:

- You must manage different sets of certificates for production, staging, and every development PC.
- You still need an HTTP server to forward invalid HTTP requests to HTTPS.
- There are subtle differences when using real and self-signed certificates. For example, browsers don't cache data from a self-signed server. Applications could run fine locally but experience cache-related issues in production.
- The Node.js app must listen on port 443 when deployed to a production server. It must be launched by a superuser (`sudo node index.js`), but this grants the app permission to do anything. *It could accidentally wipe all system files!*

---

16. https://nodejs.org/dist/latest/docs/api/fs.html#fsreadfilesyncpath-option
17. https://nodejs.org/dist/latest/docs/api/https.html#httpscreateserveroptions-requestlistener

A better approach is to use a web server such as NGINX[18] as a *reverse proxy*. It can handle SSL, HTTP requests, and static files, but forward all requests to the Node.js app (over HTTP) when necessary. (See chapter 18 for deployment options.)

Serve Static Files

Most web applications contain *static* files that return the same response to all users. These could include images, favicons, CSS stylesheets, client-side JavaScript, pre-rendered HTML pages, or any other asset.

It would be painful to programmatically assign routes for every file. Express allows you to define a single directory that contains static assets and returns any file that matches the URL path.

Create a directory named `static` in your project folder and add a file named `page.html` with the following content:

```
<!DOCTYPE html>
<html lang="en">
<head>
<meta charset="UTF-8">
<title>Static page</title>
<meta name="viewport" content="width=device-width,initial-scale=1" />
</head>
<body>

  <h1>This is a static page</h1>

</body>
</html>
```

Edit your `index.js` file and add the following code after the final `app.get()` route:

```
// serve static assets
```

18. https://www.nginx.com/

```
app.use(express.static( 'static' ));
```

(The following "Express Middleware Functions" section explains this code.)

Save and restart the application, then open http://localhost:3000/page.html in your browser.

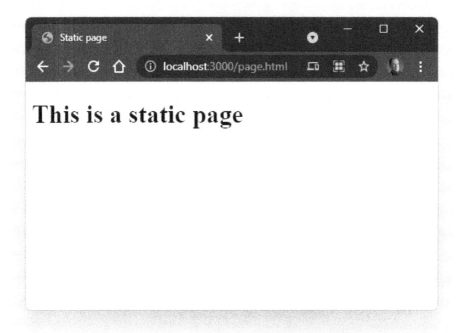

5-4. Our static page

Try adding pages, images, or other assets to the static directory or a subdirectory within it. For example, an image at /static/images/myimage.png can be viewed in the browser at http://localhost:3000/images/myimage.png .

Efficient Static Assets

In this example, Express only checks the file system for a matching static asset when it can't be handled by a routing function. However, you could check for assets first if your application mostly consists of static files.

On production servers, it's more efficient to use a frontend web server such as NGINX to serve static assets and bypass Node.js processing entirely.

Express Middleware Functions

The `app.use()` method[19] used above to define the static directory introduces the concept of Express *middleware*[20]. **Middleware** functions run in the sequence defined in the code, and can typically:

- run code on every request
- manipulate or change the request and response objects
- terminate a response—perhaps if the user isn't logged in
- call the next middleware function

In this case, `express.static( 'static' )` returns a middleware function that handles static directory processing.

All middleware functions receive three arguments:

- `req` : the Express HTTP Request object[21].
- `res` : the Express HTTP Response object[22].
- `next` : a callback that passes control to the next middleware function. Middleware functions must always call `next()` unless they complete or

---

19. https://expressjs.com/en/4x/api.html#app.use
20. http://expressjs.com/en/guide/using-middleware.html
21. https://expressjs.com/en/4x/api.html#req
22. https://expressjs.com/en/4x/api.html#res

terminate the current request.

The following middleware function logs every URL request to the terminal:

```
// log every request to the terminal
app.use((req, res, next) => {
  console.log(req.url);
  next();
});
```

You should place this function before any others that could end processing. No logging would occur if you placed it after URL routing or static asset middleware that succeeded in returning a response.

Define Working Directories

A hard-coded *static* directory is used above. That's fine for Express, but what if another module needed to locate the same directory to read or write a file?

We can define a fully qualified reference to all working directories in the *cfg* configuration object. This used to be easy in CommonJS (see Chapter 8 for more on this topic) because Node provided a *__dirname* constant[23] with the full directory of the current module. The situation is more complex in ES6 modules, because they're referenced by URL—*not by file*. The URL of the current module is available in *import.meta.url* , so it can be parsed to a file path using the standard Node.js library:

```
import { fileURLToPath } from 'url';
import { dirname, sep } from 'path';

const __dirname = dirname(fileURLToPath( import.meta.url )) + sep;
```

The *url* module[24] provides a *fileURLtoPath()* function , which converts a

---

[23]. https://nodejs.org/dist/latest/docs/api/modules.html#__dirname
[24]. https://nodejs.org/dist/latest/docs/api/url.html

`file://` URL to a fully qualified file path.

The `path` module[25] provides a `dirname()` function to extract the directory from a path and a `sep` constant with the platform-specific path separator (`/` on POSIX, `\` on Windows).

Update the top of `index.js` accordingly:

```
// Express application
import express from 'express';

import { fileURLToPath } from 'url';
import { dirname, sep } from 'path';

// configuration
const
  __dirname = dirname(fileURLToPath( import.meta.url )) + sep,
  cfg = {
    port: process.env.PORT || 3000,
    dir: {
      root:   __dirname,
      static: __dirname + 'static' + sep
    }
  };

console.dir(cfg, { depth: null, color: true });

// Express initiation
// ...rest of code
```

Then change the reference to the hard-coded `static` directory:

```
// serve static assets
app.use(express.static( cfg.dir.static ));
```

The application shows the configuration settings when starting, but the static page at http://localhost:3000/page.html should work as before.

[25.] https://nodejs.org/dist/latest/docs/api/path.html

5-5. Our Express directory configuration

Other modules can't access the *cfg* object unless you *export* it. The active *app* object can also be useful, so add the following code at the end of *index.js* :

```
// export defaults
export { cfg, app };
```

Compressing HTTP Responses

To improve web application performance, you should compress assets before they're returned to the browser over the network. The *compression*

middleware module[26] can handle this for you. Stop your app, then install the
module:

```
npm install compression
```

The *dependencies* section of your *package.json* file updates accordingly:

```
"dependencies": {
  "compression": "^1.7.4",
  "express": "^4.17.1"
}
```

Load the module at the top of *index.js* :

```
// Express application
import express from 'express';
import compression from 'compression';
```

Then add it as one of the first middleware functions (before routers and static
file handlers):

```
// HTTP compression
app.use( compression() );
```

It won't make a noticeable difference to performance here, but addressing
performance at the start of a project puts you one step ahead of most teams!

Disable Express Identification

By default, Express sets the following HTTP response header:

```
X-Powered-By: Express
```

It doesn't do any harm, but you can disable it with *app.disable()* [27] in

---

26. https://www.npmjs.com/package/compression
27. http://expressjs.com/en/4x/api.html#app.disable

index.js :

```
// Express initiation
const app = express();

// do not identify Express
app.disable('x-powered-by');
```

It will save a few bytes on every HTTP request, and will also give malicious hackers less information about your Node.js technology stack.

Handle 404 Not Found Errors

Add the following code as the *last* middleware function to gracefully handle errors when a page or asset can't be found:

```
// 404 error
app.use((req, res) => {
  res.status(404).send('Not found');
});
```

This returns a "Not Found" message with a 404 HTTP header code, but you could also do one of the following options:

- redirect[28] to an appropriate page
- show suggested pages to the user
- log bad requests to a file for further analysis

See the course *code/ch05/express02* directory[29] and associated video[30] to run the code created so far.

---

28. http://expressjs.com/en/5x/api.html#res.redirect
29. https://github.com/spbooks/ultimatenode1/tree/main/ch05/express02
30. https://spnt.co/nodevid8

Add an HTML Template Engine

Node.js has a wide range of HTML template engines that create HTML pages or snippets for output. A typical engine will take an HTML template and:

- substitute variables with actual values
- allow the inclusion of partials such as headers, footers, menus, and so on
- permit basic programming functionality, such as conditions and loops

 Template Performance

Ideally, your HTML templates should do as little as possible at runtime. You may be able to pre-render some parts of a template, such as including other files (partials) so your app has less work to do when rendering a page.

Popular templating options include Pug[31], Nunjucks[32], and EJS[33], which we'll use here, because it's one of the simplest, fastest, and most popular options. Many HTML template engines[34] work with Express, but most provide instructions in situations where there's no direct support.

In this example, you'll create a simple *message.ejs* template that's used to display single messages such as "Hello World!" in an *<h1>* tag. Stop your server and install EJS with *npm install ejs*.

The *dependencies* section of your *package.json* file updates accordingly:

```
"dependencies": {
  "compression": "^1.7.4",
  "ejs": "^3.1.6",
  "express": "^4.17.1"
}
```

[31.] https://pugjs.org/
[32.] https://mozilla.github.io/nunjucks/
[33.] https://ejs.co/
[34.] https://expressjs.com/en/resources/template-engines.html

Now create a *views* subdirectory in your project. Add a file to it named *message.ejs* with the code to output a *title* variable:

```
<%- include('partials/_htmlhead'); -%>

<h1><%= title %></h1>

<%- include('partials/_htmlfoot'); -%>
```

This template includes other partials, so create a *partials* subdirectory in *views* with a *_htmlhead.ejs* file:

```
<!DOCTYPE html>
<html lang="en">
<head>
<meta charset="UTF-8">
<title><%= title %></title>
<meta name="viewport" content="width=device-width,initial-scale=1" />
</head>
<body>
```

Also create an *_htmlfoot.ejs* file:

```
</body>
</html>
```

Open the Express entry *index.js* file and add a new *cfg.dir.views* property that points at the *views* directory:

```
// configuration
const
  __dirname = dirname(fileURLToPath( import.meta.url )) + sep,
  cfg = {
    port: process.env.PORT || 3000,
    dir: {
      root:   __dirname,
      static: __dirname + 'static' + sep,
      views:  __dirname + 'views' + sep
```

```
    }
  };
```

Add this code before any routes and middleware:

```
// use EJS templates
app.set('view engine', 'ejs');
app.set('views', cfg.dir.views);
```

This sets EJS as the Express view engine with files contained in the *views* directory.

EJS is invoked using the Express Response *render()* method[35] in a routing function. Update the functions */* , */hello/* , and the 404 handler:

```
// home page route
app.get('/', (req, res) => {
  res.render('message', { title: 'Hello World!' });
});

// another route
app.get('/hello/', (req, res) => {
  res.render('message', { title: 'Hello again!' });
});

// serve static assets
app.use(express.static( cfg.dir.static ));

// 404 errors
app.use((req, res) => {
  res.status(404).render('message', { title: 'Not found' });
});
```

The *render* method is passed the name of the template (*'message'* —the *.ejs* extension can be omitted) and an object containing name/value pairs. A *title* is set in this example.

---

35. http://expressjs.com/en/4x/api.html#res.render

Start your Express server with `npm start`, then open http://localhost:3000/ in a browser.

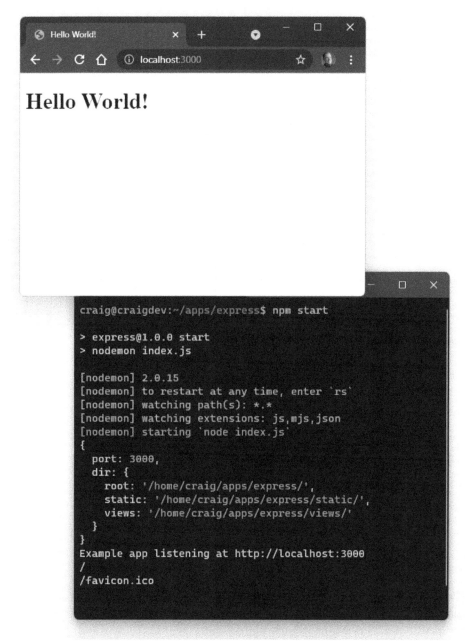

5-6. Express EJS rendering

The result may not be significantly different, but it's a fully rendered HTML page with an `<h1>` title. *(View the source, Luke.)*

Advanced Routing

URL routing is at the heart of Express processing. You've developed simple routes that run functions for specific matching URLs, but there are more options:

- path expressions: handling many routes with one function
- path parameters: parsing routes to extract values
- HTTP methods: using GET, POST, DELETE, PUT and so on
- route handlers: grouping related route handler functions into one file

Routing Path Expressions

Simple URL routes are defined in the examples above. For example:

```
// another route
app.get('/hello/', (req, res) => {
  res.send('Hello again!');
});
```

The route handles HTTP GET requests to `/hello/` , although Express will do the following:

- Ignore casing. The paths `/Hello/` and `/HELLO/` will match the `/hello/` route unless you add `app.set('case sensitive routing', true)` to `index.js` .
- Ignore closing slashes. The paths `/hello/` and `/hello` match the same route unless you add `app.set('strict routing', true)` to `index.js` .

As well as exact routes, you can define regular expression patterns to match a range of URLs. For example:

- `?` denotes that the preceding character is optional. A route of `/ab?cd/`

matches the URL paths */abcd/* and */acd/* .

- + denotes that the preceding character must appear one or more times. A route of */ab+cd/* matches the URL paths */abcd/* , */abbcd/* , */abbbbbcd/* and so on.
- * denotes any number of characters. A route of */ab*cd/* matches the URL paths */abcd/* , */ab123cd/* , */ab-node.js-cd/* and so on.
- A more complex route of */.+Script$/* matches the URL paths */JavaScript/* and */ECMAScipt/* , but not */Scripting/* .

Express uses the Path-to-RegExp[36] module to parse paths. The Express Route Tester tool[37] can help you build and debug more complex URLs.

Routing Path Parameters

Route parameters are named path segments preceded by a colon (:) to identify a variable in the URL. For example, the route */user/:id* matches any URL path starting */user/* that has a single segment—such as */user/123* or */user/abc* .

Captured values are available in the Request *params* object[38], so *req.params.id* would be set to *123* or *abc* in the examples above.

Any number of URL parameters can be defined. The following route function would run for the path */author/Craig-Buckler/book/Node.js* :

```
// return a value for a user
app.get('/author/:name/book/:bookName', (req, res, next) => {

  console.log(`author: ${ req.params.name }`);       // "Craig-Buckler"
  console.log(`  book: ${ req.params.bookName }`);  // "Node.js"

  next();
```

[36.] https://www.npmjs.com/package/path-to-regexp
[37.] http://forbeslindesay.github.io/express-route-tester/
[38.] http://expressjs.com/en/4x/api.html#req.params

```
});
```

HTTP Route Methods

The examples above handle HTTP GET requests by defining an *app.get()* function. Express supports all the other HTTP methods[39], including:

- HTTP POST with *app.post()*
- HTTP PUT with *app.put()*
- HTTP DELETE with *app.delete()*

app.all() [40] handles all HTTP methods to a specific route. The function can examine the *req.method* property to determine which HTTP method was used.

Creating a Route Handler

Defining all route handler functions in the entry *index.js* script becomes impractical as your application grows in complexity. A better option is to create route handling middleware in separate files with related functionality.

The following example updates the Express code so that requests to any URL starting */hello/* are handled in a single router file. Two GET requests are implemented:

- */hello/:name* returns a page saying hello to someone by name. For example, */hello/craig* displays "Hello Craig!"
- */hello/:lang/:name* returns a page saying hello to someone by name in a specific language. For example, */hello/fr/craig* switches to French and displays "Bonjour Craig!"

---

39. https://developer.mozilla.org/Web/HTTP/Methods
40. http://expressjs.com/en/4x/api.html#app.all

Before doing this, create a *lib* subdirectory in your project folder for generic library modules. Add a new file at *lib/locale.js* with the following code:

```
// localisation

// international greetings
export const hello = {
  au: 'G\'day',
  cn: 'Nǐ hǎo',
  en: 'Hello',
  de: 'Hallo',
  es: 'Hola',
  fr: 'Bonjour',
  jp: 'Kon\'nichiwa'
};
```

Then add *lib/string.js* with the following code:

```
// string functions

// capitalize the first letter of all words
export function capitalize(str) {

  return str
    .trim()
    .toLowerCase()
    .split(' ')
    .map(word => word.charAt(0).toUpperCase() + word.slice(1))
    .join(' ');

}
```

Next, create a new *routes* subdirectory in your project folder for routing middleware. Add a new file at *routes/hello.js* with code to define the two routing functions:

```
// /hello/ route
import { Router } from 'express';
import { hello } from '../lib/locale.js';
import { capitalize } from '../lib/string.js';
```

```
export const helloRouter = Router();

// say hello in English
helloRouter.get('/:name', (req, res, next) => {

  res.render(
    'message',
    { title: `${ hello.en } ${ capitalize( req.params.name ) }!` }
  );

});

// say hello in a specific language
helloRouter.get('/:lang/:name', (req, res, next) => {

  res.render(
    'message',
    { title: `${ hello[req.params.lang] || hello.en } ${ capitalize( req.params.
      ↪name ) }!` }
  );

});
```

This defines an Express Router object[41] named `helloRouter`. **Routers** are mini applications that can perform routing and middleware functions.

The first route defines a function for the parametrized path `/:name`. *(You should not specify the full `/hello/:name` route, because this router file will become the handler for all `/hello/` paths.)* The function renders the `message` template with a `title` that says "Hello" (in English) to the `:name` value passed on the URL (`req.params.name`).

The second route defines a function for the parametrized path `/:lang/:name`. Again, this renders the `message` template with a title that uses a localized version of "Hello" as defined in `lib/locale.js`.

To use your Router file, open `index.js` then *remove* these lines:

---

[41] http://expressjs.com/en/4x/api.html#router

```
// another route
app.get('/hello/', (req, res) => {
  res.send('Hello again!');
});
```

Replace them with this code:

```
// /hello/ route
import { helloRouter } from './routes/hello.js';
app.use('/hello', helloRouter);
```

app.use() defines the *helloRouter* middleware rather than a single
app.get() route.

If necessary, restart your Express app with *npm start* and open a URL in your
browser, such as *http://localhost:3000/hello/craig* to see "Hello Craig!"

5-7. Our hello route

Switch to an Australian greeting with the URL *http://localhost:3000/hello/*

au/craig .

5-8. Our hello route localized to Australia

See the course `code/ch05/express02` directory[42] and associated video[43] to
run the code created so far.

Exercises

Attempt the following updates to improve your Express coding experience:

▦ Improve the `message` template to add a stylesheet. *(Hint: the CSS could be
a static file.)*

▦ Create and use a new template that also outputs the current URL to the
page. *(Hint: the Express Request object passed as `req` can help.)*[44]

▦ Create a new router to say "Goodbye" in a similar way to the "Hello"
example.

Summary

This chapter introduced the Express framework for server-side web
applications. Other Node.js server frameworks follow similar conventions and
some are compatible with Express middleware.

This is just the start of the possibilities. In the following chapters, we'll look at
ways to process form data, implement REST APIs, and manipulate databases
in your Express applications.

Quiz

1. Express is:

▦ a. similar to Apache or NGINX but programmable with Node.js code
▦ b. a Node.js server-side application framework
▦ c. one of several Node.js web server frameworks
▦ d. all of the above

[42.] https://github.com/spbooks/ultimatenode1/tree/main/ch05/express03
[43.] https://spnt.co/nodevid9
[44.] http://expressjs.com/en/4x/api.html#req

2. Express is typically installed in a project as:

   a. a global module
   b. a development dependency
   c. a dependency
   d. a single static JavaScript file

3. A `package.json` file is used to:

   a. store configuration information about a Node.js application
   b. store application runtime data
   c. configure npm
   d. all of the above

4. An Express middleware function:

   a. is an internal Express module
   b. runs when an Express app starts
   c. can handle or manipulate the HTTP request and response
   d. runs when an Express app shuts down

5. Middleware functions are passed the following parameters in order:

   a. the *next* function, the Request object, the Response object
   b. the Request object, the Response object, the *next* function
   c. the *next* function, the Response object, the Request object
   d. the Response object, the Request object, the *next* function

Chapter

Processing Form Data with Express

6

Unless you're creating a static website, processing user data posted from an HTML form is at the heart of all web applications. In this chapter, you'll learn how Express can:

- parse query string data typically sent in an HTTP GET request (see the "Processing HTTP GET Query Strings" section)
- parse posted body data typically sent in an HTTP POST request (see the "Processing HTTP Post Body Data" section)
- receive uploaded files typically sent in a `multipart/form-data` HTTP POST (see the "Processing Uploaded Files" section)

 Code Examples

> The Express examples provided below purposely omit some of the options recommended in the previous chapter. Dropping features such as compression, router middleware, and 404 pages makes for more concise code—*but be sure not to forget them in your projects!*

Sanitize User Input

The rules of data processing club:

1   Never trust user data.

2   See #1.

User data must always be sanitized on the server. You may have robust HTML and JavaScript validation, but there's no guarantee the request came from a browser or worked as you expected. Always check data before it's used elsewhere—*especially if it's output to an HTML page.* (Note that the EJS `<%=` escapes HTML.)

Incoming field data will be a string, so you can check for specific formats using regular expressions[1] and parse to types such as numbers[2], dates[3], or objects[4] to check for errors. The express-validator[5] module provides a range of validation and sanitization functions.

For brevity, the examples below don't check any incoming data, so please don't use them on a live server!

Processing HTTP GET Query Strings

Data can be passed on the URL query string denoted by a `?` and a series of `name=value` pairs separated by `&` —such as `http://localhost:3000/?a=1&b=2&c=3` . Query strings are usually added to HTTP GET requests, although they can be used by any method.

---

[1] https://developer.mozilla.org/Web/JavaScript/Guide/Regular_Expressions

[2] https://developer.mozilla.org/Web/JavaScript/Reference/Global_Objects/parseFloat

[3] https://developer.mozilla.org/Web/JavaScript/Reference/Global_Objects/Date/parse

[4] https://developer.mozilla.org/Web/JavaScript/Reference/Global_Objects/JSON/stringify

[5] https://express-validator.github.io/

Express automatically parses query strings and returns a name/value object in the Request *.query* property[6]. The example URL above returns an object:

```
{
  a: 1,
  b: 2,
  c: 3
}
```

The example code in *code/ch06/express-get* [7] provides a simple example. The template *views/form.ejs* implements an HTML *<form>* , which posts to itself with its *method* set to *"get"* . A table at the top shows all name/value pairs passed in a *data* object:

```
<%- include('partials/_htmlhead'); -%>

<h1><%= title %></h1>

<% if (data) { %>
  <p>Data received in last request:</p>
  <table>
    <% for (const name in data) { %>
    <tr>
      <th>
        <%= name %>:</th>
      <td>
        <%= data[name] %>
      </td>
    </tr>   <% } %>
  </table>

<% } %>

<p>Submission form:</p>

<form action="/" method="get">
  <div>
    <label for="name">name</label>
```

---

6. http://expressjs.com/en/4x/api.html#req.query
7. https://github.com/spbooks/ultimatenode1/tree/main/ch06/express-get

```
    <input type="text" id="name" name="name" value="<%= data.name %>" />
  </div>
  <div>
    <label for="job">job</label>
    <input type="text" id="job" name="job" value="<%= data.job %>" />
  </div>
  <div>
    <label for="nodejs">like Node.js?</label>
    <input type="checkbox" id="nodejs" name="nodejs" value="yes"<% if
    ↪(data.nodejs) { %> checked<% } %> />
  </div>
  <input type="hidden" name="date" value="<%= new Date(); %>" />
  <button>submit</button>
</form>

<%- include('partials/_htmlfoot'); -%>
```

(Note that *views/partials/_htmlhead.ejs* provides a little inline CSS styling.)

The *index.js* entry script sets the EJS template engine and renders the *form* template when a GET request is made to the root / URL. The template is passed an object containing:

- the page *title*
- a *data* property set to *req.query*

```
// Express application
import express from 'express';

// configuration
const cfg = { port: process.env.PORT || 3000
};

// Express initiation
const app = express();

// use EJS templates
app.set('view engine', 'ejs');
app.set('views', 'views');
```

```
// render form
app.get('/', (req, res) => {
  res.render('form', {
    title: 'Parse HTTP GET data',
    data: req.query
  });
});

// start server
app.listen(cfg.port, () => {
  console.log(`Example app listening at http://localhost:${ cfg.port }`);
});
```

Following an *npm install* to install the Express and EJS dependencies, start the server running with *npm start* and navigate to http://localhost:3000/ in a browser.

6-1. An example form

Enter some data and hit **submit**. The URL query string changes, and all name/ value pairs are displayed. (Note that the date is passed as a *hidden* input value.)

6-2. Received data displayed with the example form

See the course `code/ch06/express-get` directory[8] and associated video[9] to run this code.

Processing HTTP Post Body Data

An HTTP POST sent via an HTML `<form>` with its `method` set to `"post"` places all data in the body of the request. Express doesn't parse this data by default and requires an `express.urlencoded()` middleware function[10] to populate a Request `.body` property[11] with an object containing name/value pairs.

8. https://github.com/spbooks/ultimatenode1/tree/main/ch06/express-get
9. https://spnt.co/nodevid10
10. http://expressjs.com/en/4x/api.html#express.urlencoded
11. http://expressjs.com/en/4x/api.html#req.body

The body-parser Module

Older editions of Express didn't include a body parsing function, so you may see references to a *body-parser* module[12] in other tutorials.

The code in *code/ch06/express-post* [13] provides a simple example. The template *views/form.ejs* is identical to that shown in the GET example above (in the "Processing HTTP GET Query Strings" section), except the *<form>* *method* is set to *"post"* .

The *index.js* entry script sets the EJS template engine and then defines the body parsing middleware like so:

```
// body parsing
app.use(express.urlencoded({ extended: true }));
```

The *extended* syntax option uses the *qs* module[14] to create a richer Request *body* object with nested properties and arrays if you've defined form fields appropriately.

The initial page load for the root / URL is an HTTP GET request, while the form submission is an HTTP POST request. Rather than define these as separate routes, the *index.js* entry script uses *app.all()* [15] so a single function processes all HTTP methods. It renders the *form* template and passes an object where the *data* property is set to *req.body* :

```
// Express application
import express from 'express';
```

---

12. https://www.npmjs.com/package/body-parser

13. https://github.com/spbooks/ultimatenode1/tree/main/ch06/express-post

14. https://www.npmjs.com/package/qs

15. http://expressjs.com/en/4x/api.html#app.all

```
// configuration
const cfg = { port: process.env.PORT || 3000
};

// Express initiation
const app = express();

// use EJS templates
app.set('view engine', 'ejs');
app.set('views', 'views');

// body parsing
app.use(express.urlencoded({ extended: true }));

// render form
// use .all to handle initial GET and POST
app.all('/', (req, res, next) => {
  if (req.method === 'GET' || req.method === 'POST') {
    res.render('form', {
      title: 'Parse HTTP POST data',
      data: req.body
    });
  }
  else {
    next();
  }
});

// start server
app.listen(cfg.port, () => { console.log(`Example app listening at
  ↪http://localhost:${ cfg.port }`);
});
```

Following an *npm install* to install the Express and EJS dependencies, start the server running with *npm start* and navigate to http://localhost:3000/ in a browser.

6-3. The example form

Enter some data and hit **submit**. The data is posted and all name/value pairs are displayed. (Note the date is passed as a *hidden* input value.)

6-4. The example form with POST data

See the course *code/ch06/express-post* directory[16] and associated video[17]
to run this code.

Processing Uploaded Files

Receiving file uploads in Express is gloriously simple compared to some
languages. However, it requires a third-party module such as *formidable* [18] to
parse incoming streamed data to one or more files.

The example code in *code/ch06/express-file* [19] has a *package.json* file
where Express, EJS, and Formidable are declared as project dependencies:

```
"dependencies": {
```

```
  "ejs": "^3.1.6",
  "express": "^4.17.1",
  "formidable": "^2.0.1"
 }
```

The template *views/form.ejs* defines a *<form>* with its *method* set to *"post"* and *enctype* set to *"multipart/form-data"* . A field that allows images to be uploaded is also added:

```
<input type="file" id="image" name="image" accept="image/*" />
```

The received data *<table>* also checks for an *imageurl* property in the *data* object and displays it using an *<img>* tag when found:

```
<table>
<% for (const name in data) { %>
  <tr>
    <th><%= name %>:</th>
    <td>
      <%= data[name] %>
      <% if (name === 'imageurl') { %>
        <img src="<%- data[name] %>" alt="uploaded image" />
      <% } %>
    </td>
  </tr>
<% } %>
</table>
```

The *index.js* entry script defines an *uploads* subdirectory, where uploaded files are stored:

```
// Express application
import express from 'express';
import formidable from 'formidable';

import { fileURLToPath } from 'url';
import { dirname, parse, sep } from 'path';

// configuration
```

```
const
  __dirname = dirname(fileURLToPath( import.meta.url )) + sep,
  cfg = {
    port: process.env.PORT || 3000,
    dir: {
      root:     __dirname,
      uploads:  __dirname + 'uploads' + sep
    }
  };
```

(Create this *uploads* subdirectory in your project. A project is somewhere
within your home directory should already have write permissions, but run
chmod 666 uploads if necessary.)

The script then initializes Express and sets *uploads* as a static directory. This
makes it easy to display an uploaded image for the purposes of this example,
but you'd normally move a valid file to a *safer* location—perhaps outside the
project directory—to ensure that it can't be accidentally deleted or
overwritten. (See the "Exercises" section below for pointers.)

```
// Express initiation
const app = express();

// use EJS templates
app.set('view engine', 'ejs');
app.set('views', 'views');

// static assets
app.use(express.static( cfg.dir.uploads ));
```

Note that the *express.urlencoded()* middleware is no longer required,
because *formidable* will also parse the form fields.

The *app.all()* [20] route uses a single function for all HTTP methods. When
this routing function runs:

▨ It initializes a new *formidable* object with the *upload* directory and a

---

[20.] http://expressjs.com/en/4x/api.html#app.all

setting to keep the file extension.

- The *.parse()* [21] method is called with the Express Request object (*req*) and a callback function that runs once the upload has completed. The callback is passed an error message (*err*), the (non-file) *data* fields, and a *files* object.
- If a single, non-empty *image* property exists in *files* , the *data* object is supplemented with information about the image. Formidable places it in the *uploads* directory with a unique GUID filename to ensure it can't clash with previous uploads.
- The *data.imageurl* property is defined by extracting the filename from the file path and prepending a slash / to define a URL that resolves to the static directory.

```
// render form
// use .all to handle initial GET and POST
app.all('/', (req, res, next) => {

  if (req.method === 'GET' || req.method === 'POST') {

    // parse uploaded file data
    const form = formidable({
      uploadDir: cfg.dir.uploads,
      keepExtensions: true
    });

    form.parse(req, (err, data, files) => {

      if (err) {
        next(err);
        return;
      }

      if (files && files.image && files.image.size > 0) {
        data.filename = files.image.originalFilename;
        data.filetype = files.image.mimetype;
        data.filesize = Math.ceil(files.image.size / 1024) + ' KB';
        data.uploadto = files.image.filepath;
        data.imageurl = '/' + parse(files.image.filepath).base;
```

21. https://github.com/node-formidable/formidable#parserequest-callback

```
      }

      res.render('form', { title: 'Parse HTTP POST file data', data });

    });

  }
  else {
    next();
  }

});

// start server
app.listen(cfg.port, () => {
  console.log(`Example app listening at http://localhost:${ cfg.port }`);
});
```

 Callback Functions

The callback function passed to *form.parse()* is the first callback example we've used. This function is called *asynchronously*: the Node.js runtime can perform other tasks while the callback waits for data.

Understanding JavaScript callbacks, promises, and *async/await* is essential for Node.js development. They're discussed further in Chapter 9.

Following an *npm install* to install the Express, EJS, and Formidable dependencies, start the server running with *npm start* and navigate to http://localhost:3000/ in a browser.

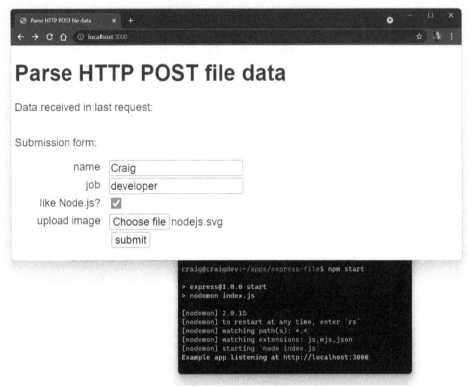

6-5. The example form

Enter some data, choose an image file, and hit **submit**. The data is posted and all name/value pairs are displayed with the image URL displayed in an `<img>` tag.

6-6. Data is displayed with the example form, along with the image

See the course `code/ch06/express-file` directory[22] and associated video[23] to run this code.

---

[22] https://github.com/spbooks/ultimatenode1/tree/main/ch06/express-file
[23] https://spnt.co/nodevid12

Exercises

Modify any of the examples so that:

- a new *email* field is added to the HTML form
- the receiving route only permits data expected in the HTML form—but nothing else
- the user values are validated—especially the email address (a basic regular expression is fine)
- adapt the EJS template to show errors as necessary

For some big bonus points, write code to delete files from the *uploads* directory—perhaps those uploaded more than 24 hours ago. You'll require Node.js file system methods[24] such as *readdir()* to read a directory, *stat()* to fetch file information, and *unlink()* to delete a file.

Summary

This chapter has built on your Express knowledge to illustrate how you can receive and process data uploaded to the server. This is essential for any web application and Express makes life a little easier for developers.

Quiz

1. Data passed on the URL query string:

- a. is not parsed in Express by default
- b. is available in an object returned by the Request *.query* property
- c. is available in an object returned by the Request *.querystring* property
- d. is available in an object returned by the Request *.body* property

2. Body data in an HTTP POST request:

- a. is not parsed in Express by default

[24.] https://nodejs.org/dist/latest/docs/api/fs.html

b. is available in an object returned by the Request `.query` property

c. is available in an object returned by the Request `.querystring` property

d. is available in an object returned by the Request `.body` property

3. File upload data in an HTTP POST request:

a. is not parsed in Express by default

b. requires a third-party module to process the incoming data

c. should be handled asynchronously in Node.js

d. all of the above

How to Use the npm Node Package Manager

7

You can attribute much of Node's success—*and frustration*—to npm. Node Package Manager provides ways to find, install, update, manage, publish, and remove Node.js packages. A **package** could be anything from a simple, one-line JavaScript module to a full application.

7-1. The npm logo

npm is the world's largest software registry. Almost 1.5 million packages have been published at <u>https://registry.npmjs.org/</u> and the majority are free to include in your own projects. You can publish your own package with a single command, and almost 1,000 developers do that every day.

Earlier chapters in this course introduced some npm concepts, but the following sections explain options you'll use daily (plus a few you'll use less frequently). The information is important, although you can skim it and use this chapter for reference later.

 npm Alternatives

npm isn't the only Node.js package manager, and you can try alternatives such as Yarn[1] and pnpm[2]. However, npm is installed with Node.js and it's good enough for most developers.

[1.] https://github.com/yarnpkg/yarn

[2.] https://github.com/pnpm/pnpm

Global vs Local Packages

By default, npm installs packages in the local project directory so it can be used in an application.

You can also install packages globally so that they're available across your whole system. This is most practical for command-line applications and utilities that could be used at any time from any directory.

For example, to install the ESLint JavaScript validator[3] globally, run `npm install eslint --global`.

You can then run `eslint <file.js>` from any directory to validate a JavaScript file.

However, you *could* install `eslint` in a project directory if you wanted to guarantee all team members had the module and fixed their errors before committing code to a project.

 npm link

> `npm link` symlinks the current project directory so it acts like a global package. A script can then be run from any other directory. This can be useful when developing a package you intend to use globally. There's no need to publish and install it as a global package every time you make a change.
>
> `npm uninstall <name> --global` removes the symlink.
>
> Don't worry if this isn't clear now. You're unlikely to use this feature until you start sharing modules with other developers.

---

3. https://eslint.org/

npm Help

npm documentation is available at **https://docs.npmjs.com/**, but help is also available from the command line: `npm help`. For further details, enter `npm help npm`, or request help about a specific npm command. For example:

```
npm help install
npm help list
npm help config
npm help package.json
```

npm Configuration

You'll rarely need to change npm configurations, but you can view your defaults with `npm config list`, or you can view a complete list of settings with `npm config list -l`.

An individual setting can be viewed. For example, show the default author name:

```
npm config get init-author-name
```

A setting can also be changed:

```
npm config set init-author-name="Craig Buckler"
```

From this point forward, npm won't prompt for the author name when initializing any project.

A setting can be unset (or deleted) with `npm config delete init-author-name`.

Project Initialization

To start a new project, you should create a new directory, navigate to it, and run `npm init`.

This prompts for information about the project—such as it name, description, Git repository, and so on. Use *npm init --yes* to accept all defaults without prompting.

npm init creates a configuration file named *package.json* . You can adapt this from another project or edit it manually if you prefer. The file contains information about your project and its dependencies. For example:

```
{
  "name": "express",
  "version": "1.0.0",
  "description": "Example Express app",
  "type": "module",
  "main": "index.js",
  "scripts": {
    "start": "nodemon index.js"
  },
  "author": "Craig Buckler",
  "license": "MIT",
  "dependencies": {
    "compression": "^1.7.4",
    "ejs": "^3.1.6",
    "express": "^4.17.1"
  }
}
```

Your project can then be installed on another device using *npm install* , which downloads all the required dependencies for the application.

Common *package.json* values include:

name	description
name	the project name—which must be unique if you want to publish on the npm registry (see the "Publishing Packages" section below)
version	the semantic version number (see the "Semantic Versioning" section below)
description	a short description of the project
type	either `"module"` for ES6 modules or `"commonjs"` (the default)
keywords	an array of strings to help others discover the project
repository	the code repository, often on GitHub
homepage	the project home page URL (often the GitHub `README.md` file)
bugs	the project issue tracker URL (often the GitHub **Issues** panel)
licence	a license for usage restrictions (if any)—set `"private"` if you're not sharing the project
main	the main entry/starting script
scripts	script commands (see the "Using npm Scripts" section below) which typically build, test, launch, or deploy a project
dependencies	project dependencies (see the "Project Dependencies" section below) required at runtime
devDependencies	development dependencies (see the "Development Dependencies" section below) required during development

Lesser-used values include:

name	description
config	application runtime configuration parameters such as ports
publishConfig	configuration parameters used at publish time
engines	the Node.js version required—such as `"node":` `">=14.0.0"`
os	an array of compatible operating systems—such as `["linux", "darwin", "win32"]`
cpu	an array of compatible CPU architectures—such as `["x64"]`
browser	the main entry/starting script for client-side JavaScript packages installed with npm (used instead of `main`)
funding	a funding page URL
files	an array of file patterns that specifies the files included when the package is installed as a dependency
bin	a list of one or more executable files to install in the PATH
man	one or more manual page files
peerDependencies	compatibility of your package with another
bundledDependencies	other packages bundled with the package
optionalDependencies	an optional dependency; the package should run without it
private	set `"true"` and npm will never publish the package to the npm registry

See the online help documentation[4] or run `npm help package.json` for a full

---

[4] https://docs.npmjs.com/configuring-npm/package-jso

description.

Semantic Versioning

Always use a semantic version for your project with `MAJOR.MINOR.PATCH` numbers separated by a period (`.`).

When a change occurs, you should increment the appropriate number and zero those that follow. Assuming a current version of `1.2.33` :

- a new bug fix would update the `PATCH` number to version `1.2.34`
- new functionality that didn't break backward compatibility would update the `MINOR` number to version `1.3.0`
- a major update with incompatible API changes would update the `MAJOR` number to version `2.0.0`

Not all developers follow this convention, so read the documentation carefully!

Project Dependencies

A package such as Express is (usually) required at runtime. It's a dependency for your application; the app would fail to run without it.

Project dependencies are listed in the `dependencies` section of `package.json` . When your project is deployed to another machine (such as a live production server), running `npm install` installs all dependencies.

Development Dependencies

Packages such as the Browsersync live reload server or the ESLint JavaScript validator are (usually) used during development. They aren't required by your application when it runs, so they aren't required on a live production server.

Development dependencies are listed in the `devDependencies` section of

package.json . They aren't installed if you run `npm install` when the
`NODE_ENV` environment variable is set to *production* . This can be set on Linux
or macOS:

```
NODE_ENV=production
```

This is the Windows *cmd* prompt:

```
set NODE_ENV=production
```

And this for Windows Powershell:

```
$env:NODE_ENV="production"
```

Searching for Packages

You'll need to install and use a third-party dependency for your application at
some point. Always consider whether you *really* need it. npm is often criticized
for reasons such as:

- There may be dozens of packages that perform a similar function. How
  long will it take to evaluate the best option?
- Installation can cause an avalanche of further installations, as each
  package requires others that have further dependences. You can even end
  up with multiple versions of the same package in the same project.
- Every third-party package and subpackage raises security implications.
  npm has a registry of known vulnerabilities, but information won't be
  available for new or less popular packages.

Is it more practical to write the code yourself?

A small module specific to your application is a good candidate. You'll learn
more and be able to write fully customizable code that's fast and lean. Over
the long term, it may even take less time and effort than maintaining a

regularly updated third-party package.

Larger or more generic modules such as frameworks (Express), database drivers, or image compressors are full projects in their own right. It makes sense to leverage the many hours of development and real-world testing.

There's an infinite array of situations between these extremes. Only you can make a judgement, but you may find yourself using fewer packages as your Node.js and JavaScript knowledge increases.

 Development Dependency Limits?

> Development tools (in `devDependencies`) have no direct effect on your application. That said, using a large number will increase installation times, require ongoing maintenance, and may confuse new team members.

Perhaps start by browsing a list of curated Node.js packages:

- https://github.com/sindresorhus/awesome-nodejs
- https://nodejs.libhunt.com/

Alternatively, you can search for packages from the command line using `npm search <term>` . For example, to find a MongoDB database driver, enter `npm search mongodb` .

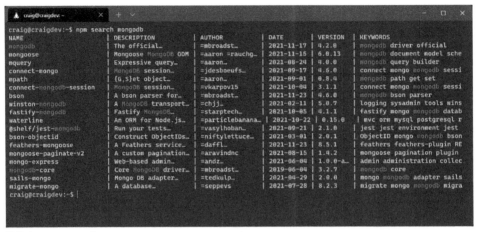

7-2. An npm search

More practically, it's best to use an online search engine:

- https://www.npmjs.com/: the official repository
- https://npms.io/: a fast search that ranks packages by quality
- https://snyk.io/advisor/: ranks packages with a health percentage

There are tools for comparing two or more packages:

- https://npmcompare.com/
- https://moiva.io/

Or tools to extract package information:

- http://npm.anvaka.com/: dependency visualization
- https://npm-stat.com/: download and usage statistics

Hardcore coders can even examine the JSON data used by npm at https://registry.npmjs.org/! Add the package name to the URL—for example, https://registry.npmjs.org/express.

If you're struggling to choose, opt for a package that's popular with a non-restrictive usage license, recent and regular updates, a small size, the fewest dependencies, and no major outstanding issues.

Installing Packages

To install a development dependency, run *npm install* , followed by one or more space-separated package names. For example:

```
npm install express mongodb
```

To install a package as a development dependency, add *--save-dev* to the command:

```
npm install browser-sync --save-dev
```

These options install the latest package into the *node_modules* directory and update *package.json* with the name and current version number.

 .gitignore node_modules

> There's no need to add the *node_modules* directory to your Git (or other) repository, because *npm install* can re-create the dependency tree.

If you require a specific or earlier package, add *@* and the version number to the package name. For example:

```
npm install ejs@2.7.4
```

To install a package globally so it's available in any directory, add *--global* to the command:

```
npm install eslint --global
```

Shortcut Aliases

Most npm commands and switches have shorter aliases. Either `i`
or `add` can be used in place of `install` , and `-g` can be used
instead of `--global` . For example:

```
npm i eslint -g
```

Semantic Constraints

`package.json` uses special codes to indicate which version of a package can
be installed on a clean machine using `MAJOR.MINOR.PATCH` semantic versioning
(see the "Semantic Versioning" section above):

- `1.2.33` : install an exact version
- `>1.2.33` : install a version greater than `1.2.33` (`2.0.0` is permitted)
- `>=1.2.33` : install a version greater than or equal to `1.2.33`
- `<1.2.33` : install a version less than `1.2.33`
- `<=1.2.33` : install a version less than or equal to `1.2.33`
- `^1.2.33` : install any greater or equal compatible version with the same
  `MAJOR` number—such as `1.3.0` but not `2.0.0` (this is the default)
- `~1.2.33` : similar to `^` but won't go beyond the next `MINOR` number—that
  is, a maximum of `1.3.0`
- `*` (or an empty string): install any version

Versions can be combined—for example, `<2.0.0 || >=3.0.0` , to skip version
`2.x.x` .

The installation of each package (and subpackage) is recorded in `package-
lock.json` . This ensures subsequent installs are identical regardless of
available updates. The file can be added to your code repository, although you
can run into problems if the application is installed on different operating
systems. Personally, I prefer to set the exact version in `package.json` , omit
`package-lock.json` from the Git repo, and then update and test manually

whenever new packages are available. (See the "Finding Outdated Packages" section below.)

"No-install" Execution

The *npx* command allows you run a package command without installing it locally. For example, try running the *cowsay* [5] talking cow package:

```
npx cowsay "I love Node.js!"
```

7-3. The cowsay package

You'll be prompted to agree to the download the first time this command is run. From then on, the version in the npm cache is used.

---

[5] https://www.npmjs.com/package/cowsay

npx Local Execution

A package such as *eslint* or *rollup* can't be run directly from the command line when it's installed locally. The following command fails if ESLint is installed locally:

```
eslint file.js
```

Rather than installing it globally, you can run a local package by defining an npm script (see the "Using npm Scripts" section below) or using npx. This command works:

```
npx eslint file.js
```

Listing Packages

To list all the packages installed in your project, enter *npm list* (or use the aliases *ls* , *la* , or *ll* in place of *list*).

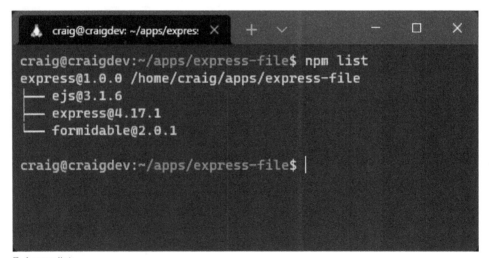

7-4. npm list

Older versions of npm show all packages and child packages. Add *--depth=0* to view the top-level installations only:

```
npm list --depth=0
```

The *--depth* argument can be used to view the package dependency tree to a specific level. For example, *npm list --depth=1* shows your installed packages and their immediate dependencies but doesn't go any deeper.

```
craig@craigdev:~/apps/express-file$ npm list --depth=1
express@1.0.0 /home/craig/apps/express-file
├── ejs@3.1.6
│   └── jake@10.8.2
├── express@4.17.1
│   ├── accepts@1.3.7
│   ├── array-flatten@1.1.1
│   ├── body-parser@1.19.0
│   ├── content-disposition@0.5.3
│   ├── content-type@1.0.4
│   ├── cookie-signature@1.0.6
│   ├── cookie@0.4.0
│   ├── debug@2.6.9
│   ├── depd@1.1.2
│   ├── encodeurl@1.0.2
│   ├── escape-html@1.0.3
│   ├── etag@1.8.1
│   ├── finalhandler@1.1.2
│   ├── fresh@0.5.2
│   ├── merge-descriptors@1.0.1
│   ├── methods@1.1.2
│   ├── on-finished@2.3.0
│   ├── parseurl@1.3.3
│   ├── path-to-regexp@0.1.7
│   ├── proxy-addr@2.0.7
│   ├── qs@6.7.0
│   ├── range-parser@1.2.1
│   ├── safe-buffer@5.1.2
│   ├── send@0.17.1
│   ├── serve-static@1.14.1
│   ├── setprototypeof@1.1.1
│   ├── statuses@1.5.0
│   ├── type-is@1.6.18
│   ├── utils-merge@1.0.1
│   └── vary@1.1.2
└── formidable@2.0.1
    ├── dezalgo@1.0.3
    ├── hexoid@1.0.0
    ├── once@1.4.0
    └── qs@6.9.3

craig@craigdev:~/apps/express-file$
```

7-5. npm list --depth=1

You can list globally installed packages using *npm list --global* .

```
craig@craigdev: ~/apps/expres:  ×    +  ∨              —    □    ×

craig@craigdev:~/apps/express-file$ npm list --global
/home/craig/.npm-global/lib
├── eslint@8.2.0
├── gulp-cli@2.3.0
├── nodemon@2.0.15
├── npm@8.1.3
└── small-static-server@1.0.3

craig@craigdev:~/apps/express-file$ |
```

7-6. npm list --global

Finding Outdated Packages

Find local packages that have received updates using *npm outdated* or global
packages with updates using *npm outdated --global* .

```
craig@craigdev: ~/apps      ×    +  ∨              —    □    ×

craig@craigdev:~/apps$ npm outdated --global
Package   Current  Wanted  Latest  Location             Depended by
eslint      8.2.0   8.3.0   8.3.0  node_modules/eslint  global
npm         8.1.3   8.1.4   8.1.4  node_modules/npm     global
craig@craigdev:~/apps$ |
```

7-7. npm outdated --global

Older packages are listed with their *current* and *latest* version. The
wanted column indicates which version will be installed if you run *npm
update* .

To update a local package, you can do one of the following:

   run *npm update* to update all packages according to semantic constraints

(see the "Semantic Constraints" section above)

- run `npm update <package>` to update one or more space-separated packages according to semantic constraints (see the "Semantic Constraints" section above)
- edit `package.json`, change any necessary version numbers, and rerun `npm install`

To update global packages, run `npm install <package> --global`. Again, any number of space-separated packages can be listed.

 Update npm with npm

> npm itself is a global package that you can update with `npm install npm --global` or the shorter `npm i npm -g`.

Removing Packages

You should always remove unused packages. They increase installation times, use disk space, could have vulnerabilities, and are likely to confuse other developers working on the project. Remove packages with `npm uninstall <package>`, (or use the aliases `remove`, `rm`, `r`, `un`, or `unlink` in place of `uninstall`).

```
craig@craigdev:~/apps/express:  ×    +   ∨              —    □    ×

craig@craigdev:~/apps/express-file$ npm list
express@1.0.0 /home/craig/apps/express-file
├── ejs@3.1.6
├── express@4.17.1
└── formidable@2.0.1

craig@craigdev:~/apps/express-file$ npm uninstall ejs

removed 15 packages, and audited 58 packages in 481ms

2 packages are looking for funding
  run `npm fund` for details

found 0 vulnerabilities
craig@craigdev:~/apps/express-file$ npm list
express@1.0.0 /home/craig/apps/express-file
├── express@4.17.1
└── formidable@2.0.1

craig@craigdev:~/apps/express-file$ |
```

7-8. npm uninstall

package.json is updated and the package is removed from the *dependencies* or *devDependencies* section. There's no need to specify the type.

Global packages can be removed with the *--global* switch. For example:

```
npm uninstall eslint --global
```

Using npm Scripts

The *"scripts"* section of *package.json* lists useful script aliases you can run during development, testing, building, deployment, and so on. A script is useful when you find yourself repeatedly retyping the same command.

Consider the JavaScript bundler Rollup[6], which can build a single optimized client-side JavaScript file[7] from multiple source files. The command to compile a development build is long. For example:

```
npx rollup --config --environment NODE_ENV:development --sourcemap --watch
  ↪--no-watch.clearScreen
```

It can therefore be defined as a script in `package.json`. For example:

```
"scripts": {
  "rollup": "rollup --config --environment NODE_ENV:development --sourcemap
  ↪--watch --no-watch.clearScreen"
}
```

Note that `npx` isn't required in the command, because npm can execute locally installed packages.

You can now start the `rollup` command with `npm run rollup`.

Any number of scripts can be added to `package.json`, but each must have a unique name.

Special Scripts

The following script names can be defined when appropriate:

- `"start"` : starts your application. You used it in previous chapters to launch `"nodemon index.js"`.
- `"test"` : runs tests on your application code using a test runner such as Mocha[8], Jest[9], or AVA[10].

---

6. https://rollupjs.org/
7. https://www.sitepoint.com/rollup-javascript-bundler-introduction/
8. https://mochajs.org/
9. https://jestjs.io/

    "stop" : stops your application. This may only be necessary if your
    application starts in the background. *I've never used it!*

The *run* command isn't required, so you can launch these scripts with *npm start* , *npm test* , and *npm stop* .

Pre and Post Scripts

Any script can have one or both of these:

    a *"pre<name>"* script, which automatically runs before *"<name>"*
    a *"post<name>"* script, which automatically runs after *"<name>"*

For example:

```
"scripts": {
  "prebuild": "rm -rf build",
  "build": "rollup --config",
  "postbuild: "echo build complete"
}
```

Running *npm run build* runs all three scripts in the order shown above.

Life Cycle Scripts

npm permits life cycle scripts[11] that automatically execute at certain points during package publication (see the "Publishing Packages" section below) or installation. The reserved script names are *prepare* , *prepublish* , *prepublishOnly* , *prepack* , and *postpack* .

You're unlikely to use these in your own projects, but avoid using these names for other purposes.

---

10. https://github.com/avajs/ava
11. https://docs.npmjs.com/using-npm/scripts

Sophisticated Scripting

npm scripts are simple but powerful. Developers often use them instead of dedicated JavaScript task runners such as Grunt[12] and Gulp[13].

Consider the following scripts to clean a build directory then generate HTML, CSS, and JavaScript using (imaginary) Node.js tools:

```
"scripts": {
  "clean"      : "rm -rf build",
  "build:html" : "sitegen ./src/content/ ./build/ --compress",
  "build:css"  : "cssgen ./src/css/main.css --out ./build/css/",
  "build:js"   : "jsgen ./src/js/main.js ./build/js/main.js --minify"
}
```

A single *build* script could run the *clean* script followed by all build tools in parallel:

```
  "build"      : "clean && (build:html & build:css & build:js)"
```

Executing *npm run build* performs all tasks in a bash shell. However, it won't work in Windows or other shells that don't support *&* and *&&* command chaining.

Cross-platform scripts can be created using task packages such as yall-scripts[14], concurrently[15], or npm-run-all[16]. The rimraf[17] package can also replace the *rm* command.

You can install cross-platform modules:

---

[12] https://gruntjs.com/
[13] https://gulpjs.com/
[14] https://www.npmjs.com/package/yall-scripts
[15] https://www.npmjs.com/package/concurrently
[16] https://www.npmjs.com/package/npm-run-all
[17] https://www.npmjs.com/package/rimraf

```
npm install yall-scripts rimraf --save-dev
```

Then update *package.json* to use them:

```
"scripts": {
  "clean"      : "rimraf build",
  "build:html" : "sitegen ./src/content/ ./build/ --compress",
  "build:css"  : "cssgen ./src/css/main.css --out ./build/css/",
  "build:js"   : "jsgen ./src/js/main.js ./build/js/main.js --minify",
  "buildcode"  : "yall --parallel build:*",
  "build"      : "yall --sequential clean buildcode"
}
```

npm run build will now work on any platform that can run Node.js.

Publishing Packages

Your own packages can be published to the npm repository. This may be
practical when you want to share code with others or create your own libraries
for use in several projects. Skip down to the "Exercises" section if you'd rather
think about this later!

 Publication Preparation

> Publishing code to the npm repository makes it public. Always
> ensure it doesn't contain private information such as Git or
> database credentials.

> Authors of popular packages receive regular requests for support
> or feature updates. Add a disclaimer to the *README.md* file in the
> root of your project if you'd rather not offer a free consultancy
> service! That said, you can request funding[18] and watch the cash
> roll in as your package becomes an essential part of every Node.js
> project.

[18.] https://docs.npmjs.com/cli/v8/configuring-npm/package-json#funding

To publish a package, you must sign up for an account at npmjs.com. A valid email address is required, and it will be publicly added to the metadata of any package you publish.

 Two-factor Authentication

> Accounts are secured with 2FA, so you'll need an app such as Google Authenticator, Microsoft Authenticator, Authy, or andOTP.

Before publishing, update your package `package.json` file:

1   Use a unique `"name"`.

All npm projects must have a unique name. Naming is difficult. You have 1.5 million competitors, so use a tool such as the npm-package-name-checker[19] to check availability. If you can't find a decent name, prefix the name with your account ID—such as `@username/my-package`.

2   Set the next semantic `"version"` number (See the "Semantic Versioning" section above).

You can't *overwrite* an existing package with the same version number. The next unique version must be set every time you publish.

3   Add an optional array of `"files"` glob patterns[20].

You can define which files are included in the package. The following example includes all files and subdirectories in the `dist` and `doc` directories. All other files except `package.json` are omitted:

```
"files": [
  "dist/**/*",
```

---

19. https://remarkablemark.org/npm-package-name-checker/
20. https://en.wikipedia.org/wiki/Glob_(programming)

```
  "doc/**/*"
],
```

4   Add optional *"bin"* command(s)[21] aliases.

To run your package by its *package.json* *"name"*, define a relative path to its script as a *"bin"* value. For example:

```
"name": "myapp",
"bin": "./dist/myapp.js"
```

The *myapp* command can be run from the command line when the package is installed globally. (*npx myapp* can be used for local installations.)

"bin" can also be set to an array if you have more than one script or require aliases. For example:

```
"name": "myapp",
"bin": [
  "myapp": "./dist/myapp.js",
  "ma": "./dist/myapp.js",
  "myapp2": "./dist/myapp2.js"
]
```

To publish, navigate to your project directory, then log in at the terminal with *npm login*. Publish your package with *npm publish*.

Assuming there are no errors, npm will publish your package so it can be installed from anywhere. At this point, it's best to commit the code to your repository to ensure the codebases are the same.

[21.] https://docs.npmjs.com/cli/v8/configuring-npm/package-json#bin

Publishing Tips

You're unlikely to publish many packages at the start of your Node.js journey, but the following tips may help as you develop more complex projects:

- Create packages that meet your needs to solve a specific problem.
- Create small, focused packages that do one thing well and can be reused across many projects.
- It may be better to create a new package than complicate an existing one.

In summary: **keep it simple.**

Exercises

Attempt the following exercise to improve your npm knowledge:

1   Initialize a new Node.js project, ideally using a name that's not already taken in the npm registry.

2   Search for packages that can output colors to the terminal.

3   Install your chosen package into the project.

4   Create a small command-line application that's passed a string and color argument. Output the string in that color.

5   Optionally, publish the code to npm, then install it as a global package so you can run it from anywhere.

The video for this chapter[22] describes a solution that's available in the example code[23], the npm registry[24], and GitHub[25].

---

[22] https://spnt.co/nodevid13
[23] https://github.com/spbooks/ultimatenode1/tree/main/ch07/concol
[24] https://www.npmjs.com/package/concol
[25] https://github.com/craigbuckler/concol

Summary

This chapter has expanded on your npm knowledge so you can find, install, update, manage, publish, and remove Node.js packages in any project.

The next chapter looks at the options for using these packages and your own modules in Node.js applications.

Quiz

1. npm help is available from:

   a. online documentation
   b. the `npm help` command
   c. using `npm help <command>`
   d. any of the above

2. A Node.js `package.json` file can be initialized with:

   a. `npm new`
   b. `npm init`
   c. `npm start`
   d. any of the above

3. Your project's `package.json "version"` is currently `"1.2.33"` and you are adding a new feature (it won't break backward compatibility). The new version number should be:

   a. `2.0.0`
   b. `1.3.0`
   c. `1.3.1`
   d. `1.2.34`

4. How do you install a package for use in your project?

a. *npm add <name> --local*
b. *npm require <name>*
c. *npm install <name>*
d. any of the above

5. How do you list all the packages installed in your project without viewing any child dependencies?

a. *npm list*
b. *npm ll*
c. *npm ls --depth=0*
d. any of the above

6. How can you find packages that have newer updates in the local project?

a. *npm outdated*
b. *npm old*
c. *npm newer*
d. *npm update*

Using ES2015 and CommonJS Modules

8

The previous chapter explained how npm can be used to find and install packages containing multiple JavaScript files, or **modules**. In this chapter, we'll examine how modules are used in Node.js.

 Skip Ahead?

The information in this chapter is important, since you'll encounter issues with older Node.js packages. However, all the packages referenced in this course have been tested for compatibility, so you can skip ahead and return when you eventually run into a problem!

Modules provide a way to define functionality in one file and use it in another. Developers often create encapsulated code libraries responsible for handling related tasks. The benefits include:

- code can be split into smaller files with self-contained functionality
- the same modules can be shared and reused across any number of applications
- modules need never be examined or updated by others once they've been proven to work
- code referencing a module understands it's a required dependency
- modules prevent naming conflicts: function `x()` in `module1.js` can't clash with function `x()` in `module2.js`

Bizarrely, there was no concept of modules in JavaScript during its first twenty years. You couldn't directly reference or include one JavaScript file in another. Client-side developers would either:

- add multiple `<script>` tags to an HTML page
- concatenate scripts into a single file, perhaps using a bundler such as webpack or task runners such as Grunt and Gulp
- use a module loading library such as RequireJS[1] or SystemJS[2]—all of which adopted syntaxes such as CommonJS[3], AMD[4], or UMD[5]

[1.] http://requirejs.org/

[2.] https://github.com/systemjs/systemjs

It would have been inconceivable for Node.js not to support modules when it was released in 2009. CommonJS syntax was chosen as the Node.js module *standard*, and support was added to npm.

CommonJS

A CommonJS module makes a function or value publicly available using `module.exports` . For example:

```
// lib.js
const PI = 3.1415926;

// add values
function sum(...args) {
  log('sum', args);
  return args.reduce((num, tot) => tot + num);
}

// multiply values
function mult(...args) {
  log('mult', args);
  return args.reduce((num, tot) => tot * num);
}

// private logging function
function log(...msg) {
  console.log(...msg);
}

module.exports = { PI, sum, mult };
```

A *require* statement includes a module by referencing either:

- its relative file path (*./lib.js* , *../lib.js*)
- a fully qualified file path (*/path/lib.js*)
- its npm name following installation (*express* , *chalk* , etc.)

---

3. http://www.commonjs.org/
4. https://github.com/amdjs/amdjs-api/wiki/AMD
5. https://github.com/umdjs/umd

The module is included at the point it's referenced during execution of the script.

You can *require* specific named exported items:

```
const { sum, mult } = require('./lib.js');

console.log( sum(1,2,3,4) );   // 10
console.log( mult(1,2,3,4) );  // 24
```

Or you can *require* all exported items using a (namespaced) variable:

```
const lib = require('./lib.js');

console.log( lib.PI );            // 3.1415926
console.log( lib.add(1,2,3,4) );  // 10
console.log( lib.mult(1,2,3,4) ); // 24
```

A module with a single exported item can be defined as a default. For example:

```
// myclass.js
class MyClass {}
module.exports = MyClass;
```

And it can be defined using any name:

```
const
  MyNewClass = require('myclass.js'),
  myObj = new MyNewClass();
```

CommonJS dynamically imports file names by default, and can also import JSON data as a JavaScript object. For example:

```
const
  file = `data${ Math.round(Math.random() * 3) }.json`,
  data = require(file);
```

```
console.log( data.propertyOne || 'propertyOne not set' );
```

However, top-level *await* isn't supported. Asynchronous calls must be wrapped in an **immediately invoked function expression** (IIFE)—*a function that runs as soon as it's defined.* For example:

```
function waitOneSec() {
  return new Promise(
    (resolve) => setTimeout(resolve, 1000)
  );
}

(async () => {
  await waitOneSec();
})();
```

CommonJS was the Node.js module *standard* until the arrival of ES2015 modules.

ES2015 Modules (ESM)

A native JavaScript module standard was proposed in ES2015 (ES6).

Everything inside an ES2015 module is private by default and runs in strict mode (there's no need for `'use strict'`). Public properties, functions, and classes are exposed using `export` . For example:

```
// lib.js
export const PI = 3.1415926;

// add values
export function sum(...args) {
  log('sum', args);
  return args.reduce((num, tot) => tot + num);
}

// multiply values
export function mult(...args) {
```

```
    log('mult', args);
    return args.reduce((num, tot) => tot * num);
  }

  // private logging function
  function log(...msg) {
    console.log(...msg);
  }
```

Alternatively, a single *export* statement can declare one or more public items. For example:

```
  // lib.js
  const PI = 3.1415926;

  // add values
  function sum(...args) {
    log('sum', args);
    return args.reduce((num, tot) => tot + num);
  }

  // multiply values
  function mult(...args) {
    log('mult', args);
    return args.reduce((num, tot) => tot * num);
  }

  // private logging function
  function log(...msg) {
    console.log(...msg);
  }

  export { PI, sum, mult };
```

An *import* statement includes ES modules using either:

- a relative URL (starting `./` or `../`)
- a fully qualified URL (such as *file:///home/path/lib.js*)
- its npm name following installation (*express* , *chalk* , etc.)

Importing External URLs

Deno and browser JavaScript can import URLs from other domains:

```
import { someFunction } from 'https://example.com/lib.js';
```

This isn't supported in Node.js but will arrive in a future release. You can use an HTTPS loader[6], although it's slower than disk access, the module isn't cached, and there are security implications.

All ES modules and their submodules are resolved and imported once *before your script executes. It doesn't matter where they're declared in your script.*

You can *import* specific named items:

```
import { sum, mult } from './lib.js';

console.log( sum(1,2,3,4) );  // 10
console.log( mult(1,2,3,4) ); // 24
```

Or imports can be aliased to resolve naming conflicts:

```
import { sum as addAll, mult as multiplyAll } from './lib.js';

console.log( addAll(1,2,3,4) );      // 10
console.log( multiplyAll(1,2,3,4) ); // 24
```

Alternatively, all public items can be imported into a namespaced variable:

```
import * as lib from './lib.js';

console.log( lib.PI );            // 3.1415926
console.log( lib.add(1,2,3,4) );  // 10
console.log( lib.mult(1,2,3,4) ); // 24
```

---

[6] https://nodejs.org/dist/latest/docs/api/esm.html#https-loader

A module with a single item to export can set a *default*. For example:

```
// moduleWithDefault.js
export default function() { ... };
```

Or:

```
// moduleWithDefault.js
function x() { ... };
export default x;
```

The default is imported without curly braces and can use any name. For example:

```
import myDefault from './moduleWithDefault.js';
```

This is effectively the same as this:

```
import { default as myDefault } from './moduleWithDefault.js';
```

Dynamic module loading—perhaps from a generated value—is possible using the *import()* function, which returns a promise. For example:

```
const
  script = `./script${ Math.round(Math.random() * 3) }.js`
  randomImport = await import(script);
```

This affects performance and makes code validation difficult. Only use the *import()* function when there's no other option—for example, an imported script is created after the application starts.

Node.js version 17 and above also support JSON loading and parsing using the *import()* function:

```
import data from './data.json' assert { type: 'json' };
```

Finally, ESM supports top-level *await* . For example:

```
function waitOneSec() {
  return new Promise(
    (resolve) => setTimeout(resolve, 1000)
  );
}

await waitOneSec();
```

Comparison of CommonJS and ES2015 Modules

CommonJS and ES2015 module syntaxes are superficially similar, but they work in different ways:

- Each CommonJS *require* references a file that's dynamically loaded on demand during execution.
- Each ESM *import* references a URL that's hoisted and pre-parsed to resolve further imports. This occurs before your code is executed. Dynamic importing of modules isn't directly supported or recommended.

Consider this ES2015 module:

```
// ESM two.mjs
console.log('running two');
export const hello = 'Hello from two';
```

It's imported by this script:

```
// ESM one.mjs
console.log('running one');
import { hello } from './two.mjs';
console.log(hello);
```

This is the output when running *node one.mjs* :

```
running two
running one
hello from two
```

Now consider this CommonJS module:

```
// CommonJS two.cjs
console.log('running two');
module.exports = 'Hello from two';
```

It's required by this script:

```
// CommonJS one.cjs
console.log('running one');
const hello = require('./two.cjs');
console.log(hello);
```

This is the output when running `node one.cjs` :

```
running one
running two
hello from two
```

Execution order is critical in some applications—*and what would happen if ES2015 and CommonJS modules were mixed in the same file?*

It took several years for ESM support to arrive in Node.js. The following approach was adopted to resolve potential compatibility problems:

- CommonJS is the default (or set `"type": "commonjs"` in `package.json`).
- Any file with a `.cjs` extension is parsed as CommonJS.
- Any file with a `.mjs` extension is parsed as ESM.
- Running `node --input-type=module index.js` parses the entry script as ESM.
- Setting `"type": "module"` in `package.json` parses the entry script as ESM.

Importing CommonJS Modules in ES2015

Node.js can *import* a CommonJS module into an ESM file. For example:

```
import lib from './lib.cjs';
```

This *usually* works well, and Node.js makes syntax suggestions when problems occur.

Requiring ES2015 Modules in CommonJS

You *can't* *require* an ES module in a CommonJS file. ESM modules load asynchronously, so they aren't compatible with synchronous loading and execution in CommonJS.

One way around this is the dynamic *import()* function[7], which loads a module on demand:

```
// CommonJS script
(async () => {

  const lib = await import('./lib.mjs');

  // ... use lib ...

})();
```

Alternatively, the *esm* package[8] provides a way to import ESM code in CommonJS.

This chapter's video[9] demonstrates how CommonJS and ESM modules can be used interchangeably.

---

[7] https://developer.mozilla.org/docs/Web/JavaScript/Reference/Statements/import#dynamic_import
[8] https://www.npmjs.com/package/esm
[9] https://spnt.co/nodevid14

Using ES2015 Modules in Browsers

This section isn't specific to Node.js, but it may be useful if you're developing a cross-platform JavaScript library that works both client-side and server-side (it's *isomorphic*).

Browsers load ES modules asynchronously and defer execution until the DOM is ready. They run in the order specified by each `<script>` tag:

```
<script type="module" src="./runsfirst.js"></script>
<script type="module" src="./runssecond.js"></script>
```

Or as specified by an inline `import` :

```
<script type="module">
import { something } from './somewhere.js';
// ...
</script>
```

Browsers *without* ESM support don't load scripts with a `type="module"` attribute. Browsers *with* ESM support don't load scripts with a `nomodule` attribute:

```
<script type="module" src="runs-when-ESM-supported.js"></script>
<script nomodule src="runs-when-ESM-is-not-supported.js"></script>
```

Modules must be served with the MIME type `application/javascript` or `text/javascript` . A CORS header[10] such as `Access-Control-Allow-Origin: *` must also be set if a module can be imported from another domain.

Summary

The module situation in Node.js can be confusing. It has reached a point where:

---

10. https://developer.mozilla.org/Web/HTTP/CORS

- some libraries are CommonJS
- some libraries are ESM
- some libraries provide builds for both CommonJS and ESM

CommonJS was the only option for several years. There's little benefit converting a large project to ESM, especially where it uses older modules with compatibly issues.

Moving forward, ES2015 module syntax is the JavaScript standard implemented in browsers and the Deno runtime[11]. Personally, I like CommonJS, but I recommend ES modules for new Node.js projects. All the examples in this course use ESM. Importing CommonJS modules into ESM is usually possible, but you may need to consider alternative packages if problems occur.

For more information, refer to:

- JavaScript modules on MDN[12]
- CommonJS modules on nodejs.org[13]
- ECMAScript modules on nodejs.org[14]

Quiz

1. Node.js natively supports the following module syntaxes:

- a. CommonJS and ECMAScript modules
- b. CommonJS and AMD
- c. ECMAScript modules and UMD
- d. AMD and UMD

2. Which syntax does CommonJS use to declare and use public module functions?

---

[11]. https://deno.land/
[12]. https://developer.mozilla.org/docs/Web/JavaScript/Guide/Modules
[13]. https://nodejs.org/dist/latest/docs/api/modules.html
[14]. https://nodejs.org/dist/latest/docs/api/esm.html

a. *export* and *import*
b. *module.exports* and *import*
c. *module.exports* and *require*
d. *export* and *require*

3. Which syntax do ES modules use to declare and use public functions?

a. *export* and *import*
b. *module.exports* and *import*
c. *module.exports* and *require*
d. *export* and *require*

4. Which of the following is true?

a. CommonJS and ESM operate identically
b. you can usually *import* a CommonJS modules in ESM
c. you can usually *require* an ES module in CommonJS
d. all of the above

5. The *import()* function:

a. can import an ES module into CommonJS
b. can dynamically load an ES module after the application starts
c. returns a promise
d. all of the above

Chapter

Asynchronous Programming in Node.js

9

This chapter discusses the benefits and challenges of asynchronous programming in JavaScript. Asynchronous concepts are rarely evident in other languages, but it's impossible to avoid them in Node.js.

You may have written asynchronous event handling functions in client-side JavaScript. These should run quickly, and pages don't remain open for long; a bug could cause problems for an individual user, but a browser restart or page reload would fix it. However, your Node.js app is the central point of access for *all* users and must remain active without a restart. A small asynchronous bug can generate memory leaks that eventually crash the application.

This is one of the biggest causes of confusion when developers migrate from other languages, so *please don't skip this chapter!* Asynchronous programming can seem complex, but a few pointers will help you avoid common pitfalls.

Single-threaded Non-blocking I/O Event-looping What?

Imagine you're running a pizza restaurant on your own. You take all the orders and prepare all the pizzas but can only manage one task at a time. You receive your first order, then prepare the dough (20 minutes), add the toppings (20 minutes), pop it in the oven, watch while it cooks (20 minutes), and serve to the customer. The process takes one hour; you're then free to take another order.

9-1. A single-chef restaurant

To make your restaurant more efficient, you hire three chefs: one to make dough, one to add toppings, and one to bake. The chefs are in different kitchens and can't talk to each other, but they'll report back to you when their specific task is complete.

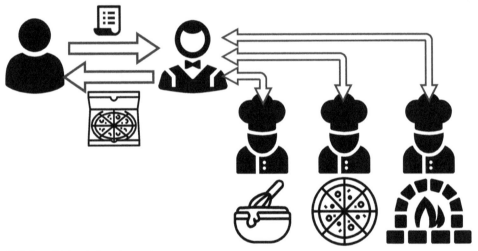

9-2. A multiple-chef restaurant

It still takes an hour to create one pizza (although the three chefs together can prepare three pizzas every hour). What's important is that you're no longer involved in the cooking process. You're passing instructions to chefs and receiving an alert when they've completed their job. You're free to take customer orders whenever they arrive.

Both JavaScript and Node.js are **single-threaded**: the runtime can only do one thing at a time. It would be like a restaurant with a single chef, except JavaScript offloads input and output operations to the operating system kernel (other "chefs" who operate in parallel). A Node.js app may start a file write, database read, or HTTP request, but it won't wait for that operation to finish. Instead, it asks for a callback function to be run when it's complete and success or error data is available.

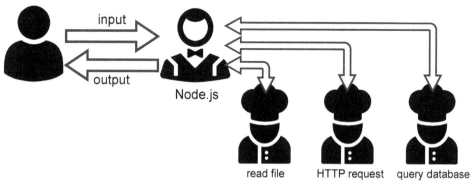

9-3. Node.js I/O

Callbacks in Action

Consider this PHP code to write text to a file:

```php
<?php
echo 'saving file';
$err = file_put_contents('file.txt', 'Hello from PHP');
if ($err !== false) echo 'file saved';
echo 'processing complete';
?>
```

The program outputs this:

```
saving file
file saved
processing complete
```

The PHP interpreter processes the *file_put_contents() statement* and waits until the file is fully written before progressing to the next command.

This is the equivalent code in Node.js:

```
import { writeFile } from 'fs';

console.log('saving file');
writeFile('file.txt', 'Hello Node.js', 'utf8', err => {
  if (!err) console.log('file saved');
});
console.log('processing complete');
```

The program outputs this:

```
saving file
processing complete
file saved
```

Processing completes *before* the file saves!

The fourth argument passed to `writeFile()` [1] is an anonymous ES6 callback function with a single `err` parameter. The callback runs when the file has saved (or fails to save and raises an error passed in `err`). File saving may only take a few milliseconds, but it runs in the background, so the `'processing complete'` command executes immediately.

Asynchronous callbacks are at the heart of all client-side and server-side JavaScript applications.

It's standard practice to handle errors and return an error object or string message as the *first* argument to a callback function (like `err` above). When no error occurs, the callback's first parameter should be `null` , `undefined` or any other falsy value.

---

[1] https://nodejs.org/dist/latest/docs/api/fs.html#fswritefilefile-data-options-callback

The Event Loop

Why does the Node.js program above continue to run after the last line has executed?

All Node.js applications initialize an event loop. Once the last statement completes execution, Node.js loops back and checks for any outstanding:

- timers (such as `setTimeout`)
- pending callbacks
- polling data connections

These are run in the order they're received (know as "first in, first out", or FIFO).

A seemingly idle application won't end if it's waiting for something to complete or if something could occur at a future point (such as a server `listen` on a specific port).

 Avoid Blocking the Event Loop

Long-running JavaScript calculations or processes block the event loop and delay the processing of incoming requests. Process-intensive tasks should either be:

- split into smaller sub-tasks with timers
- run in the background using a worker thread or a child process (options that are discussed in Chapter 12).

Callback Conundrums

Using callbacks in asynchronous functions isn't always easy. Your code can look correct and run without errors, but it eventually causes the Node.js runtime to crash.

Two severe issues are:

- failing to terminate an asynchronous function after a callback
- accidentally making an asynchronous function synchronous

These are best explained with examples. Consider this simple asynchronous function which waits for *ms* milliseconds:

```
// wait for ms milliseconds
function wait(ms, callback) {

  setTimeout(callback, ms);

}

// wait for one second
wait(1000, () => {
  console.log('waited 1000ms');
});
```

Let's improve the function by returning the following arguments to the callback:

- an error when *ms* is not a number, less than 1, or more than 3000
- the value of *ms* waited

Our initial implementation:

```
// wait for ms milliseconds
function wait(ms, callback) {

  ms = parseFloat(ms);

  // invalid ms value?
  if (!ms || ms < 1 || ms > 3000) {

    const err = new RangeError('Invalid ms value');
    callback( err, ms );
```

```
  }

  // wait ms before callback
  setTimeout( callback, ms, null, ms );

}

// call wait
wait(500, (err, ms) => {

  if (err) console.log(err);
  else console.log(`waited ${ ms }ms`);

});
```

Execution returns the expected `waited 500ms` result.

9-4. Wait callback—500ms

However, what happens when we pass an invalid `ms` value such as `0` ?

```
craig@craigdev: ~/apps/callbac  ×    +  ∨                        —    □    ✕
craig@craigdev:~/apps/callback$ node callback-fail.js
RangeError: Invalid ms value
    at wait (/home/craig/apps/callback/callback-fail.js:9:17)
    at Object.<anonymous> (/home/craig/apps/callback/callback-fail.js:21:1)
    at Module._compile (node:internal/modules/cjs/loader:1101:14)
    at Object.Module._extensions..js (node:internal/modules/cjs/loader:1153:10)

    at Module.load (node:internal/modules/cjs/loader:981:32)
    at Function.Module._load (node:internal/modules/cjs/loader:822:12)    at Fu
nction.executeUserEntryPoint [as runMain] (node:internal/modules/run_main:81:12
)
    at node:internal/main/run_main_module:17:47
waited 0ms
craig@craigdev:~/apps/callback$ |
```

9-5. Wait callback—0ms

We get the error we expected, but the *setTimeout* also runs and we see
waited 0ms . The callback function executes twice, because the function
didn't terminate when the error occurred. We can solve this by putting the
setTimeout in an *else* statement or adding a *return* in the error condition:

```
// wait for ms milliseconds
function wait(ms, callback) {

  ms = parseFloat(ms);

  // invalid ms value?
  if (!ms || ms < 1 || ms > 3000) {

    const err = new RangeError('Invalid ms value');
    callback( err, ms );
    return; // terminate function

  }

  // wait ms before callback
  setTimeout( callback, ms, null, ms );

}
```

There's another, subtler issue: *the callback runs immediately when an error is
raised.* At that point, the function is no longer asynchronous—*it's synchronous.*

It won't cause an obvious problem here, but it can lead to memory leaks in larger, long-running Node.js applications. Your app will eventually crash with an obscure "memory overflow" error message.

 ## A Function Must be 100% Synchronous or 100% Asynchronous

> No path through an asynchronous function should ever lead to a callback being run immediately.

A simple way to solve this is the `setImmediate()` timer[2]. This calls a function during the next iteration of the event loop:

```
// wait for ms milliseconds
function wait(ms, callback) {

  ms = parseFloat(ms);

  // invalid ms value?
  if (!ms || ms < 1 || ms > 3000) {

    const err = new RangeError('Invalid ms value');
    setImmediate( callback, err, ms );
    return;

  }

  // wait ms before callback
  setTimeout( callback, ms, null, ms );

}
```

---

2. https://nodejs.org/dist/latest/docs/api/timers.html#setimmediatecallback-arg

process.nextTick()

> You may see *process.nextTick(callback)* [3] used in some
> applications. This works similarly to *setImmediate()* , except that
> the callback runs *before* the end of the current iteration of the
> event loop. This can cause the event loop to never restart if
> *nextTick()* is recursively called.

Callback Hell

In complex Node.js applications, you'll often make a series of asynchronous
function calls—such as when fetching something from a database, making an
API call, loading a file, and so on. A callback may be used in one place only, so it
makes sense to declare an inline anonymous function. This can quickly
descend into deeply nested callback hell:

```
wait(100, (err) => {

  console.log('wait 1');

  wait(200, (err) => {

    console.log('wait 2');

    wait(300, (err) => {
      console.log('wait 3');
    });

  });
});
```

There are syntactical ways to flatten this structure, typically by naming each
function and ensuring each calls others in turn. Fortunately, the JavaScript
gods addressed the problem with promises.

---

[3] https://nodejs.org/dist/latest/docs/api/process.html#processnexttickcallback-args

Promises

A *Promise* object represents the eventual completion or failure of an asynchronous operation with its resulting value. Promises provide a clearer syntax that makes it easier to chain asynchronous calls that run in series. Developers can also avoid the callback issues raised in the previous sections.

Promises were introduced in ES6/2015 and are syntactical sugar; callbacks are still used under the hood. To make a function asynchronous, a *Promise* object[4] must be returned immediately. The *Promise* constructor is passed two callback functions as parameters:

- *resolve* : the function that's run when processing successfully completes
- *reject* : the function that's run when an error occurs

In the case of our *wait()* function, it can be rewritten to return a promise that calls *resolve(ms)* after the timeout or *reject(error)* when an invalid *ms* parameter is passed:

```
// wait for ms milliseconds
function pWait(ms) {

  ms = parseFloat(ms);

  return new Promise((resolve, reject) => {

    if (!ms || ms < 1 || ms > 3000) {
      reject( new RangeError('Invalid ms value') );
    }
    else {
      setTimeout( resolve, ms, ms );
    }

  });
```

---

[4] https://developer.mozilla.org/en-US/docs/Web/JavaScript/Reference/Global_Objects/Promise

```
}
```

 util.promisify()

> $util.promisify()$ [5] converts any callback-based function with an
> error as the first argument into a promise. Rather than re-writing
> $wait()$, you could create a promisified alternative named
> $pWait()$:

```
import { promisify } from 'util';
const pWait = promisify(wait);
```

Anything that returns a promise can have:

- a $then()$ method, which is passed a function that takes the result from the previous $resolve()$
- a $catch()$ method, which is passed a function that runs when an error is returned from any $reject()$
- a $finally()$ method, which is called at the end regardless

```
pWait(100)
  .then(ms => console.log(`waited ${ ms }ms`) );
  .catch(err => console.log( err ) )
  .finally(() => console.log('all done') )
```

Each `.then()` function can return a value or another promise so that
sequential asynchronous function calls can be chained. For example:

```
pWait(100)
  .then(ms => {
    console.log(`waited ${ ms }ms`);
    return pWait(ms + 100);
  })
```

[5.] https://nodejs.org/dist/latest/docs/api/util.html#utilpromisifyoriginal

```
.then(ms => {
  console.log(`waited ${ ms }ms`);
  return pWait(ms + 100);
})
.then(ms => {
  console.log(`waited ${ ms }ms`);
})
.catch(err => {
  console.log( err );
});
```

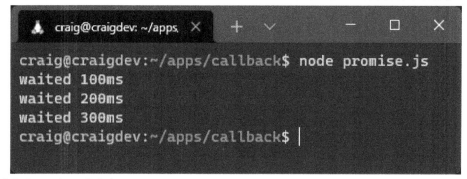

9-6. A promise call chain

 then() Functions Are Promisified

The final *then()* in the code above runs a synchronous function, but JavaScript automatically converts it into a promise-based asynchronous function so you can append further *then()* methods when necessary.

Parallel Promises

The example above executes each asynchronous function call one after the other. This is only necessary if the result from one function is required as input for the next.

You'll often encounter situations when several asynchronous functions are required but they aren't related to each other. For example, given a book ID, such as an ISBN, you want to:

- retrieve book information such as the title, author, etc. from a local database (*getBook(ID)*)
- call a stock control system API to determine how may of those books are available (*getStock(ID)*)
- get the latest recommended retail price from the publisher (*getPrice(ID)*)

Assume each function returns a promise where *resolve()* returns an information object.

The following promise chain works but is inefficient, because each call occurs one after the other:

```
// book data object
const bookData = { id: 123 };

getBook( bookData.id )

  .then(book => {
    bookData.title = book.title;
    bookData.author = book.author;
    bookData.description = book.description;
    getStock( bookData.id );
  })

  .then(stock => {
    bookData.stock = stock;
    getPrice( bookData.id );
  })

  .then(price => {
    bookData.price = price;
  })

  .catch(err => {
    console.log( err );
  })
```

A better option is *Promise.all()* [6], which takes an array of promises, runs each in parallel, and returns a new *outer* promise where `resolve()` returns an array of output values in the same order. This code is as fast as the slowest function:

```
// book data object
const bookData = { id: 123 };

Promise.all([
    getBook( bookData.id ),
    getStock( bookData.id ),
    getPrice( bookData.id )
  ])

  .then(result => {

    bookData.title = result[0].title;
    bookData.author = result[0].author;
    bookData.description = result[0].description;
    bookData.stock = result[1];
    bookData.price = result[2];

  })

  .catch(err => {
    console.log( err );
  })
```

The `.catch()` is triggered whenever a single promise `reject()` runs, so any pending promises are aborted.

Similar options include:

`Promise.allSettled()` [7]

Runs all promises in the array and waits until every one has resolved or

---

6. https://developer.mozilla.org/docs/Web/JavaScript/Reference/Global_Objects/Promise/all
7. https://developer.mozilla.org/docs/Web/JavaScript/Reference/Global_Objects/Promise/allSettled

rejected. Each item in the returned array is an object with a `.status` property (either `'fulfilled'` or `'rejected'`) and a `.value` property with the returned value.

`Promise.any()` [8]

Runs all promises in the array but resolves as soon as the first promise resolves. A single value is returned.

`Promise.race()` [9]

Runs all promises in the array but resolves or rejects as soon as the first promise resolves or rejects. A single value is returned.

Promising Problems

Promises help prevent callback hell, but I found them confusing at first, and it's easy to mangle the `.then()` / `.catch()` chain syntax. You should also note that the whole promise chain is asynchronous, so any function using a series of promises should return its own promise (or it could run a callback to confuse other developers!)

async/await

ES2017 introduced the `async` and `await` keywords, which enable asynchronous, promise-based behavior to be written in a cleaner and clearer syntax. Again, they're more syntactical sugar, but they make promises sweeter.

A promise chain to make three successive `pWait()` calls is long and difficult to read:

[8] https://developer.mozilla.org/docs/Web/JavaScript/Reference/Global_Objects/Promise/any
[9] https://developer.mozilla.org/docs/Web/JavaScript/Reference/Global_Objects/Promise/race

```
pWait(100)
  .then(ms => {
    console.log(`waited ${ ms }ms`);
    return pWait(ms + 100);
  })
  .then(ms => {
    console.log(`waited ${ ms }ms`);
    return pWait(ms + 100);
  })
  .then(ms => {
    console.log(`waited ${ ms }ms`);
  })
  .catch(err => {
    console.log( err );
  });
```

This is the equivalent code using *await* :

```
try {

  const p1 = await pWait(100);
  console.log(`waited ${ p1 }ms`);

  const p2 = await pWait(p1 + 100);
  console.log(`waited ${ p2 }ms`);

  const p3 = await pWait(p2 + 100);
  console.log(`waited ${ p3 }ms`);

}
catch(err) {
  console.log(err);
}
```

Put the *await* keyword before any promise-based asynchronous function
and the JavaScript interpreter will appear to *wait* until it's resolved or rejected.
The syntax is cleaner and looks much like a series of synchronous function
calls.

The code above is a top-level *await* because it's not contained in a function.
This works in ES2015 modules, but not in CommonJS, where you must wrap it

in an asynchronous immediately invoked function expression (IIFE):

```
(async () => {

  try {

    const p1 = await pWait(100);
    console.log(`waited ${ p1 }ms`);

    const p2 = await pWait(p1 + 100);
    console.log(`waited ${ p2 }ms`);

    const p3 = await pWait(p2 + 100);
    console.log(`waited ${ p3 }ms`);

  }
  catch(err) {
    console.log(err);
  }

})();
```

Any function that contains one or more *await* statements must have *async* prepended to indicate it's asynchronous. In effect, this turns it into a promise-based function:

```
// async function
async function waitSeries(ms) {

  try {

    const p1 = await pWait(ms);
    console.log(`waited ${ p1 }ms`);

    const p2 = await pWait(p1 + 100);
    console.log(`waited ${ p2 }ms`);

    const p3 = await pWait(p2 + 100);
    console.log(`waited ${ p3 }ms`);

  }
```

```
  catch(err) {
    console.log(err);
  }

}

// top-level await to run the async function
await waitSeries(100);
```

Promise.all() is Still Necessary

There's no *async* / *await* equivalent for *Promise.all()* and similar functions. However, *async* functions return a promise, so they can be passed in the processing array.

try/catch is Ugly

async functions silently exit if you omit *try* / *catch* and the current *await* is rejected. Unless you can examine the error type, it's not possible to know which *await* triggered the problem, so multiple *try* / *catch* blocks may be necessary.

You could consider using a higher-order function to catch errors when they can be processed in the same way. For example:

```
// async function
async function waitSeries(ms) {

  const p1 = await pWait(ms);
  console.log(`waited ${ p1 }ms`);

  const p2 = await pWait(p1 + 100);
  console.log(`waited ${ p2 }ms`);

  const p3 = await pWait(p2 + 100);
  console.log(`waited ${ p3 }ms`);
}
```

```
// higher-order function handle errors
function catchErrors(fn) {
  return function(...args) {
    return fn(...args).catch(err => {
      console.log('ERROR', err);
    });
  }
}

// top-level await
await catchErrors(waitSeries)(100);
```

Whether this results in more readable code is another matter.

Asynchronous Awaits in Synchronous Loops

Be wary about using *await* in looping methods such as *forEach()*, which are passed a function. Loops are synchronous and continue to run even when the function they call is asynchronous. Consider this example:

```
const ms = [100, 200, 300];
let totalWait = 0;

ms.forEach(async i => {

  console.log(i);
  const w = await pWait(i);
  console.log(`waited ${ w }ms`);
  totalWait += w;
});

console.log(`total wait time: ${ totalWait }ms`);
```

You might expect to see the following output:

```
100
waited 100ms
200
waited 200ms
300
```

```
waited 300ms
total wait time: 600ms
```

The actual result is surprising, as pictured below.

9-7. The await loop output

Each iteration of the loop won't *await* until it's complete. This will be a problem if the result of one *await* is required in the next call.

Standard *for()* , *while()* and async iterator loops may be necessary. The code above can be fixed with this:

```
const ms = [100, 200, 300];
let totalWait = 0;

for (let i = 0; i < ms.length; i++) {

  console.log( ms[i] );
  const w = await pWait( ms[i] );
  console.log(`waited ${ w }ms`);
  totalWait += w;
}

console.log(`total wait time: ${ totalWait }ms`);
```

The result is pictured below.

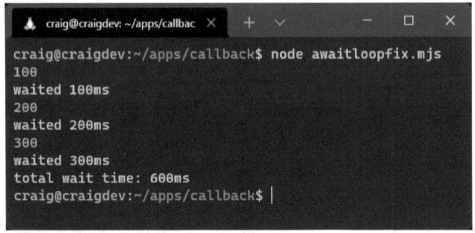

9-8. The await loop fix

Exercises

Write a small application that fetches three random questions from the Open Trivia Database[10] using the following REST URL requests:

▓ General knowledge:

   https://opentdb.com/api.php?type=multiple&amount=1&category=9

▓ Computers:

   https://opentdb.com/api.php?type=multiple&amount=1&category=18

▓ Gadgets:

   https://opentdb.com/api.php?type=multiple&amount=1&category=30

---

10. https://opentdb.com/

💡 HTTP Requests

Unless you're using Node.js 18 or above, which offers a native `Fetch()` API, you'll need use a third-party HTTP request module such as `node-fetch` [11].

Format the question data into a single array and output it in JSON format into a file named `questions.json`.

For bonus points, make your application more efficient by running all URL requests in parallel.

This chapter's video[12] demonstrates a solution.

Summary

Asynchronous programming takes some time to understand and will catch you out. The following tips will help you write more robust Node.js applications.

- Ensure JavaScript functions run quickly and don't block the event loop.
- Pass callback functions to an asynchronous function so they can be called when an operation is complete.
- The first argument of the callback function must be an error object or string message.
- Always ensure a function `return` occurs after a callback runs.
- An asynchronous function must be 100% asynchronous: no path should lead to an immediate callback. Pass a callback to `setImmediate()` to run it during the next iteration of the event loop if necessary.
- Learn how to create your own promise functions or create them from a callback-based function using `util.promisify()`.
- You can `await` for a promise to complete inside an `async` function.

---

11. https://www.npmjs.com/package/node-fetch
12. https://spnt.co/nodevid15

- Where possible, run promises in parallel using options such as
  `Promise.all()` or `Promise.allSettled()`.

Useful links:

- The Node.js event loop[13]
- Don't block the event loop (or the worker pool)[14]
- MDN promise documentation[15]
- MDN `async`[16] and `await`[17]
- The "Promises" and "Async functions" sections of Chapter 11, *JavaScript: Novice to Ninja*[18]
- Chapters 8 and 9 of *JavaScript: The New Toys*[19]

Quiz

1. A callback function:

- a. runs before an operation starts
- b. calls an asynchronous function
- c. is called when an asynchronous operation completes
- d. all of the above

2. An asynchronous function:

- a. completes at a later time
- b. allows subsequent JavaScript commands to be executed
- c. can be implemented with callbacks, promises, or `async`

---

[13] https://nodejs.org/en/docs/guides/event-loop-timers-and-nexttick/

[14] https://nodejs.org/en/docs/guides/dont-block-the-event-loop/

[15] https://developer.mozilla.org/docs/Web/JavaScript/Reference/Global_Objects/Promise

[16] https://developer.mozilla.org/docs/Web/JavaScript/Reference/Statements/async_function

[17] https://developer.mozilla.org/docs/Web/JavaScript/Reference/Operators/await

[18] https://www.sitepoint.com/premium/books/javascript-novice-to-ninja-2nd-edition

[19] https://www.sitepoint.com/premium/books/javascript-the-new-toys

d. all of the above

3. The Node.js event loop:

a. reruns when there are outstanding timers or callbacks
b. runs asynchronous functions
c. is another name for callbacks
d. none of the above

4. A *Promise* object completes by running:

a. a *resolve* or *reject* function
b. a *fulfilled* or *error* function
c. a *resolve* or *error* function
d. a *fulfilled* or *reject* function

5. An *async* function:

a. can call promise-based functions using *await*
b. returns a promise
c. uses *try* / *catch* blocks to handle errors
d. all of the above

Chapter

Using Database Storage

10

The previous chapters explained programming practices that affect all Node.js applications. This chapter applies these fundamentals to the specific challenge of data storage using database solutions such as MongoDB and MySQL.

Web applications often require data that persists between page loads and application restarts. Consider a content management system such as WordPress: it stores articles, metadata, media, user profiles, comments, settings, plugin configurations, and more. Multiple users can log in at any time to view and update content.

The most common solution to data persistence is a database such as MongoDB[1], MySQL[2], or PostgreSQL[3]. All database systems have the same purpose: *to provide the ability to store and query data fast and frequently.* They differ in how they achieve those goals.

 Skip Ahead?

Databases may not be the most exciting topic, but it's one of the most significant differences between frontend and backend engineering. You can skip sections about specific systems, but the following chapters will be more difficult to understand without some database knowledge.

A Database-driven Web Application Example

The sections below explain how to create a web page hit counter service. Your grandparents will tell you how popular they were in the 1990s.

---

[1] https://www.mongodb.com/
[2] https://www.mysql.com/
[3] https://www.postgresql.org/

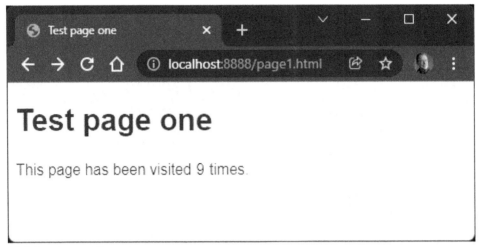

10-1. A web page hit counter

A page using this service includes an image:

```
<img src="http://localhost:8001/hit.svg" alt="hits" referrerpolicy="unsafe-url" />
```

(Note that the `referrerpolicy` is required in modern browsers so they send referral information in the HTTP header.)

The image is returned from a Node.js Express application, which:

1   Extracts the page URL from the request's `referer` HTTP header.

2   Cleans and transforms the URL to a 32-character hash using the MD5 algorithm[4]. All URLs therefore resolve to a 32-character string no matter how long they are. *(In theory, two different URLs could generate the same hash, but it's unlikely to occur for a few billion years.)*

3   Stores the hash in a new database record with the user's IP address, user agent string, and current date/time.

---

[4] https://en.wikipedia.org/wiki/MD5

4   Counts all references to the hash in the database.

5   Generates and returns an SVG image with that count.

Three applications are provided in the code directory[5]>:

1   A MongoDB version (see the "MongoDB" section below) using the native *mongodb* [6] driver.

2   A MySQL version (see the "MySQL" section below) using the native *mysql2* [7] driver.

3   A Sequelize ORM version (see the "Sequelize ORM" section below). This also connects to a MySQL database using *mysql2* [8], but you don't use it directly.

All three use the same Node.js code except for a `lib/pagehit.js` file, which communicates with a specific database to add and query records.

It's impossible to describe every option in every database, but this example code provides a head start when developing your own applications.

Installing and Configuring Database Software

You can download, install, and configure MySQL, MongoDB, or any other database on Linux, macOS, and Windows. That's beyond the scope of this course, so prepare yourself for several hours of effort.

An easier option is Docker[9]. This is often shrouded in mystery, but Docker

---

5. https://github.com/spbooks/ultimatenode1/tree/main/ch10
6. https://www.npmjs.com/package/mongodb
7. https://www.npmjs.com/package/mysql2
8. https://www.npmjs.com/package/mysql2
9. https://www.docker.com/

provides a way to download, install, and configure pre-built applications in minutes. Install Docker[10] on your system, then follow the steps below to run MySQL, MongoDB, and the Adminer database client[11]. The page hit service runs as a Node.js application on your device that connects to this database.

MongoDB

MongoDB[12] is a popular NoSQL database that became associated with Node.js in the same way MySQL is often paired with PHP. MongoDB groups JSON-like documents into one or more collections (analogous to tables) and implements querying with JavaScript-like objects.

NoSQL has become a catch-all term for any database that doesn't follow SQL conventions (see the "MySQL" section below). In general, NoSQL databases implement fewer rules. Repeated (denormalized) data is encouraged, and there's no need to define data structures, defaults, constraints, or relationships.

NoSQL software and storage mechanisms vary. Some offer basic key–value pairs. Some use JSON documents. Others are use-case specific, such as Redis[13] for in-memory caching, and Elasticsearch[14] for search-engine indexing.

A NoSQL database can be practical when data is more organic and relationships are looser. Consider an address book storing telephone numbers for individual contacts:

- You *could* allocate a single `telephone` field in an SQL database, but it's too restrictive: contacts may have home, work, and mobile numbers. Allocating three telephone fields would be wasteful for some contacts, but not enough for others. A separate `telephone` table is the most flexible option,

---

[10.] https://dockerwebdev.com/tutorials/install-docker/
[11.] https://www.adminer.org/
[12.] https://www.mongodb.com/
[13.] https://redis.io/
[14.] https://www.elastic.co/

but this increases complexity.

In a NoSQL database, telephone numbers can be defined as an unlimited array of objects associated with a contact. For example:

```
{
  "firstName": "Contact",
  "LastName": "One",
  "telephone": [
      { "home": "1-01234567890" },
      { "work": "2-01234567890" },
      { "iPhone": "3-01234567890" },
      { "Android phone": "4-01234567890" },
      { "Test phone": "5-01234567890" }
  ]
}
```

Start the MongoDB Application

To use the MongoDB-based application, navigate to the `pagehit-mongodb` directory and start MongoDB and the Adminer client with `docker-compose up` .

 Your Own MongoDB Installation?

Database configuration parameters are defined in the project's `.env` file. It configures Docker, and the Node.js application reads it using the < `dotenv` [15] module.

If you're using your own installation of MongoDB, edit the `.env` file and change the configuration parameters accordingly. In most cases, only the root user's password need be changed (`MONGO_INITDB_ROOT_PASSWORD`).

In another terminal, install the Node.js `express` , `mongodb` , and `dotenv`

[15.] https://www.npmjs.com/package/dotenv

dependencies referenced in *package.json* :

```
npm install
```

Then start the page hit application:

```
npm start
```

Finally, start a web server in another terminal so you can load test pages:

```
npx small-static-server 8888 ./test
```

You now have four services running:

- the MySQL database at *http://localhost:3306*
- the Adminer database client at *http://localhost:8080/*
- the page hit service at *http://localhost:8001/*
- a test page web server at *http://localhost:8888/*

Different ports can be defined in the *.env* file if you have clashes.

Visit http://localhost:8888/page1.html or http://localhost:8888/page2.html to view page counters. Refresh and watch the counter increase.

You can examine the database data using the Adminer panels at http://localhost:8080/. Log on with the credentials specified in *.env* :

- System: **MongoDB**
- Server: **host.docker.internal** (or your network IP address if not using Docker Desktop)
- Username: **root**
- Password: **rootuserpw**
- Database: **pagehitmongo**

10-2. Adminer login

Click the `hit` collection followed by **Select data**.

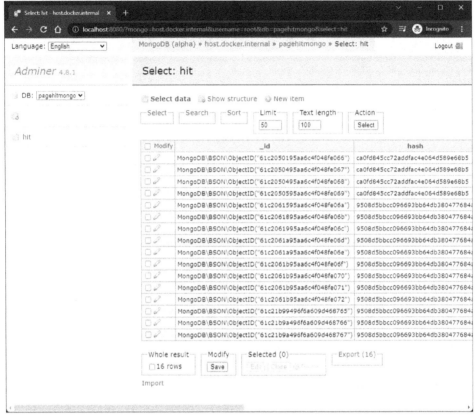

10-3. Adminer data

View this chapter's video[16] to see the code in action.

MongoDB Functionality

The `lib/pagehit.js` file handles all MongoDB functionality. It loads the required modules and extracts the configuration parameters from the `.env` file using the `dotenv` [17] module:

```
import dotenv from 'dotenv';
import { MongoClient } from 'mongodb';
```

16. https://spnt.co/nodevid16
17. https://www.npmjs.com/package/dotenv

```
import httpReferrer from './httpreferrer.js';

// load .env configuration
dotenv.config();
```

You require a Node.js package to communicate with a database. These are often referred to as database **clients**, **connectors**, or **drivers**, and the MongoDB native driver[18] is used here. It provides low-level methods to construct and execute any MongoDB command.

A connection string is passed to the *MongoClient* driver constructor, which sets the database user's name, password, host, and port. The asynchronous *.connect()* method is called to establish a connection:

```
// connect to MongoDB
const client = new MongoClient(
  `mongodb://${ process.env.MONGO_INITDB_ROOT_USERNAME }:${ process.env.MONGO_
  ↪INITDB_ROOT_PASSWORD }@${ process.env.MONGO_INITDB_HOST }:${ process.env.
  ↪MONGO_INITDB_PORT }/`,
  { useNewUrlParser: true, useUnifiedTopology: true }
);

await client.connect();
```

The code then connects to a specific database (*pagehitmongo*) and references a *hit* collection for later use (a **collection** is a group of similar JSON-like documents):

```
const
  db = client.db(process.env.MONGO_INITDB_DATABASE),
  hit = db.collection('hit');
```

MongoDB allows you to arbitrarily add data to a document in a collection without describing that data up front (although it's possible to define a schema so you can benefit from data validation). However, you should index regularly queried values to make searches faster and more efficient.

[18.] https://www.npmjs.com/package/mongodb

What Is a Database Index?

An index is a list of the data in one or more fields in a specific order—much like the index in a book. For example, you could have a number of *user* records created as each person registers. When someone logs in, you must locate a user's record by their email address:

- Without an index, the database must search through every *user* record one by one until the correct email is found.
- With an index on the *email* field in ascending alphabetical order, the database can locate a matching record far faster.

Indexes should be used on fields that you frequently query. It's tempting to create indexes on every field, but the more you add, the longer it takes to write new records and update all indexes.

The *hit* collection has an index created on the URL *hash* and *time* . This runs every time the application starts, but is ignored after the first attempt:

```
// add collection index
await hit.createIndex({ hash: 1, time: 1 });
```

lib/pagehit.js exports a single default asynchronous function. It generates a *hash* from the referring page's URL, but returns *null* when no referrer is found:

```
// count handler
export default async function(req) {

  // hash of referring URL
  const hash = httpReferrer(req);

  // no referrer?
  if (!hash) return null;
```

The browser's IP address (*ip*), user agent (*ua*), and access time (*time*) are then determined:

```
// fetch browser IP address and user agent
const
  ipRe  = req.ip.match(/(?:\d{1,3}\.){3}\d{1,3}/),
  ip    = ipRe?.[0] || null,
  ua    = req.get('User-Agent') || null,
  time  = new Date();
```

This data is added as a new document into the *hit* collection using the *insertOne()* method[19]. By default, all MongoDB documents also have a unique *_id* added to every document:

```
try {

  // store page hit
  await hit.insertOne({ hash, ip, ua, time });
```

A count of all documents with the same *hash* is then returned:

```
// fetch page hit count
return await hit.countDocuments({ hash });
```

An error is thrown if any database operation fails:

```
  }
  catch (err) {
    throw new Error('DB error', { cause: err });
  }

}
```

The main *index.js* script loads this module:

```
import pagehit from './lib/pagehit.js';
```

[19.] https://docs.mongodb.com/manual/reference/method/db.collection.insertOne/

It uses it within a middleware function that sets *req.count* to the returned page count. This is available to subsequent (*next()*) middleware functions, but any error terminates the request immediately:

```
// page hit count middleware
app.use(async (req, res, next) => {

  try {
    req.count = await pagehit(req);

    if (req.count) {
      next();
    }
    else {
      res.status(400).send('No referrer');
    }

  }
  catch(err) {
    res.status(503).send('Pagehit service down');
  }

});
```

A single */hit.svg* route is defined, which returns an SVG image containing the *req.count* value:

```
// SVG counter response
app.get('/hit.svg', (req, res) => {

  res
    .set('Content-Type', 'image/svg+xml')
    .send(`<svg xmlns="http://www.w3.org/2000/svg" width="${ String( req.count ).
    ↪length * 0.6 }em" height="1em"><text x="50%" y="75%" font-family=
    ↪"sans-serif" font-size="1em" text-anchor="middle" dominant-baseline=
    ↪"middle">${ req.count }</text></svg>`);

});
```

The response ends once the SVG is returned to the calling browser.

Stop the MongoDB Application

Stop both the Node.js page hit application and test page server by pressing `Ctrl` | `Cmd` + `C` in their terminals. From the same project directory, stop the MongoDB database and Adminer client with `docker-compose down` .

MySQL

MySQL[20] is a popular SQL database. SQL (Structured Query Language) is a standard for managing data in a relational database management system (RDBMS). Data is stored in tables and should ideally be defined in one place without duplication (known as **normalization**).

Consider a book store inventory. Each book has an ID, title, author, and publisher, and is added as a new row (record) to a `book` table:

id	title	author	publisher
1	Introduction to Node.js	Craig Buckler	SitePoint
2	Jump Start Web Performance	Craig Buckler	SitePoint
3	DevTool Secrets	Craig Buckler	SitePoint
4	Learn to Code with JavaScript	Darren Jones	SitePoint

An author and publisher can have more than one book. Rather than repeat the same values, it's more practical to create `author` and `publisher` tables where each record has a unique ID.

Here's the `author` table:

id	name	country
2	Craig Buckler	UK
3	Darren Jones	UK

[20.] https://www.mysql.com/

Here's the *publisher* table:

id	name	country
1	SitePoint	AU

You can reference those IDs in the *book* table:

id	title	author_id	publisher_id
1	Introduction to Node.js	2	1
2	Jump Start Web Performance	2	1
3	DevTool Secrets	2	1
4	Learn to Code with JavaScript	3	1

If a publisher changes their name or address, you can update the data in the *publisher* table without affecting related *book* records.

A brief overview of SQL:

- Database table structures must be defined before data can be stored.
- SQL offers simple declarative CRUD operations such as `INSERT`, `SELECT`, `UPDATE`, and `DELETE`, but is powerful enough for complex operations.
- Queries can `JOIN` tables to examine related data in a single command.
- Data integrity and relationships can be enforced. For example, it becomes impossible to delete an author if they have one or more books.
- Most systems can wrap multiple updates into a single transaction. If one operation fails, the data rolls back to the state before the first update.
- SQL is a fairly loose standard. Similar syntaxes are implemented across relational database management systems, but features and syntax can differ.
- SQL was initially devised in the early 1970s, so software, tools, documentation, and resources are plentiful.

Other popular SQL databases include MariaDB[21], PostgreSQL[22], SQLite[23], Microsoft SQL Server[24], and Oracle[25].

Start the MySQL Application

To use the MySQL-based application, navigate to the *pagehit-mysql* directory and start MySQL and the Adminer client with *docker-compose up* .

 Your Own MySQL Installation?

As before, database configuration parameters are defined in the project *.env* file, which you can edit if you're using your own MySQL installation.

Docker automatically runs the *mysql/init.sql* script to initialize the database tables and indexes. You must run this manually before starting the Node.js application, either by running it in a MySQL client or using the terminal command:

```
mysql -h localhost -u pagehituser pagehitmysql < mysql/init.sql
```

(Change the host, user, or database names as necessary.)

In another terminal, install the Node.js *express* , *mysql2* , and *dotenv* dependencies referenced in *package.json* :

```
npm install
```

Then start the page hit application:

[21.] https://mariadb.org/

[22.] https://www.postgresql.org/

[23.] https://www.sqlite.org/

[24.] https://www.microsoft.com/sql-server/

[25.] https://www.oracle.com/database/

```
npm start
```

Finally, start a web server in another terminal so you can load test pages:

```
npx small-static-server 8888 ./test
```

You now have four services running:

- the MySQL database at
- the Adminer database client at http://localhost:8080/
- the page hit service at http://localhost:8001/
- a test page web server at http://localhost:8888/

Different ports can be defined in the `.env` file if you have clashes.

Visit http://localhost:8888/page1.html or http://localhost:8888/page2.html to view page counters. Refresh and watch the counter increase.

You can examine the database data using the Adminer panels at http://localhost:8080/. Log on with the credentials specified in `.env`:

- System: **MySQL**
- Server: **host.docker.internal** (or your network IP address if not using Docker Desktop)
- Username: **pagehituser**
- Password: **pagehitpw**
- Database: **pagehitmysql**

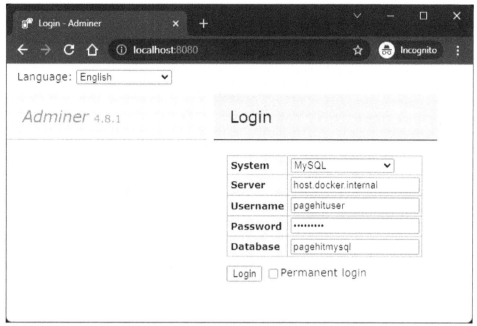

10-4. Adminer login

Then click **select** next to the `hit` table.

10-5. Adminer data

MySQL Functionality

You can't store data in an SQL RDBMS until the data structure (its **schema**) is defined. The MySQL database schema is defined in *mysql/init.sql* , which runs automatically when using Docker:

```
-- MySQL database initialization
USE pagehitmysql;

CREATE TABLE IF NOT EXISTS hit (
  id bigint unsigned NOT NULL AUTO_INCREMENT COMMENT 'record ID',
  hash binary(16) NOT NULL COMMENT 'URL hash',
  ip int(4) unsigned DEFAULT NULL COMMENT 'client IP address',
  ua varchar(200) DEFAULT NULL COMMENT 'client useragent string',
  time timestamp NOT NULL DEFAULT CURRENT_TIMESTAMP COMMENT 'hit time',
  PRIMARY KEY (id),
  KEY hash_time (hash, time)
) ENGINE=InnoDB DEFAULT CHARSET=utf8 COMMENT='page hits';
```

The *lib/pagehit.js* file handles all MySQL functionality. It loads the required modules and extracts the configuration parameters from the *.env* file using the *dotenv* [26] module:

```
import dotenv from 'dotenv';
import mysqlPromise from 'mysql2/promise';
import httpReferrer from './httpreferrer.js';

// load .env configuration
dotenv.config();
```

The *mysql2* [27] driver has been chosen for MySQL communication. It provides promise-based, low-level methods to construct and execute any SQL command.

A MySQL connection pool is configured using defaults from the *.env* file. Connection pools reduce the time spent connecting to a MySQL server by reusing previous connections:

```
// connect to MySQL
const db = await mysqlPromise.createPool({
  host:     process.env.MYSQL_HOST,
  port:     process.env.MYSQL_PORT,
  database: process.env.MYSQL_DATABASE,
  user:     process.env.MYSQL_USER,
  password: process.env.MYSQL_PASSWORD,
  waitForConnections: true,
  connectionLimit: 10,
  queueLimit: 0
});
```

Like before, *lib/pagehit.js* exports a single default asynchronous function. It generates a *hash* from the referring page's URL, returns when no referrer is found, and determines the browser's IP address (*ip*), and user agent (*ua*):

---

26. https://www.npmjs.com/package/dotenv
27. https://www.npmjs.com/package/mysql2

```
// count handler
export default async function(req) {

  // hash of referring URL
  const hash = httpReferrer(req);

  // no referrer?
  if (!hash) return null;

  // fetch browser IP address and user agent
  const
    ipRe  = req.ip.match(/(?:\d{1,3}\.){3}\d{1,3}/),
    ip    = ipRe?.[0] || null,
    ua    = req.get('User-Agent') || null;
```

 No Time?

> The time of record insertion is automatically handled by MySQL,
> which sets the `CURRENT_TIMESTAMP` by default.

The data is added as a new record into the `hit` table by executing an SQL
INSERT statement[28]:

```
try {

  // store page hit
  await db.execute(
    'INSERT INTO `hit` (hash, ip, ua) VALUES (UNHEX(?), INET_ATON(?), ?);',
    [ hash, ip, ua ]
  );
```

This is an example of a **prepared statement**, where each `?` character is
substituted by an associated (and escaped) value in the array.

---

28. https://dev.mysql.com/doc/refman/en/insert.html

UNHEX? INET_ANON?

A couple of MySQL-specific functions are used in the SQL statement above to make smaller, more efficient numeric fields that use less space and are quicker to search:

- `UNHEX()` [29] converts the 32-character hash string to a binary value.
- `INET_ATON()` [30] converts a dotted-quad IPv4 network address string to an integer.

Never Build SQL Strings!

Never programmatically build SQL strings. This is the biggest cause of SQL injection attacks:

```
b.execute(`SELECT * FROM user WHERE email='${ email }' AND
  ↳password='${ password }';`);
```

A user could enter the email address: *boss@company.com'; --* . This comments out the password check, so anyone can log in as the boss!

A more dangerous example would wipe the *user* table:

```
boss@company.com'; DROP TABLE user; --
```

You should validate all incoming user data, but a prepared statement makes SQL injection attacks far more difficult.

A count of all records with the same *hash* is then returned:

---

29. https://dev.mysql.com/doc/refman/en/string-functions.html#function_unhex
30. https://dev.mysql.com/doc/refman/8.0/en/miscellaneous-functions.html#function_inet-aton

```
// fetch page hit count
const [res] = await db.query(
  'SELECT COUNT(1) AS `count` FROM `hit` WHERE hash = UNHEX(?);',
  [ hash ]
);

return res?.[0]?.count;
```

An error is thrown if any database operation fails:

```
  }
  catch (err) {
    throw new Error('DB error', { cause: err });
  }

}
```

As before, the main `index.js` script loads the `lib/pagehit.js` module, sets `req.count` in a middleware function, and outputs it in a generated SVG in the `/hit.svg` route.

Stop the MySQL Application

Stop both the Node.js page hit application and test page server by pressing `Ctrl`|`Cmd` + `C` in their terminals. From the same project directory, stop the MongoDB database and Adminer client with `docker-compose down`.

Sequelize ORM

The MySQL and MongoDB examples in their respective sections above use native drivers to communicate directly with a database using its SQL or NoSQL command syntax. This has some disadvantages:

- Your application is tied to a specific database.
- You must learn and implement the language used by that database.
- You must track your own data and schema updates to ensure database changes are pushed to all installations of the application.

An object-relational mapping (ORM) module can make development easier by providing an abstract layer between your code and the database. Rather than running SQL/NoSQL commands directly, your code manipulates data objects that are saved and restored from a representation in a database.

`sequelize` [31] is a popular Node.js ORM that supports MySQL, MariaDB, PostgreSQL, SQLite, SQL Server, and other SQL databases. It still requires a native database driver such as `mysql2`, but there's no need to write SQL statements.

Start the Sequelize ORM Application

To use the ORM-based application, navigate to the `pagehit-orm` directory and start MySQL and the Adminer client with `docker-compose up`.

 Your Own MySQL Installation?

> As before, database configuration parameters are defined in the project `.env` file, which you can edit if you're using your own MySQL installation. In this case, there's no initialization script, because it's handled by Node.js code.

In another terminal, install the Node.js `express`, `sequelize`, `mysql2`, and `dotenv` dependencies referenced in `package.json`:

```
npm install
```

Then start the page hit application:

```
npm start
```

Finally, start a web server in another terminal so you can load test pages:

---

31. "https://www.npmjs.com/package/sequelize

```
npx small-static-server 8888 ./test
```

You now have four services running:

- the MySQL database at `http://localhost:3306`
- the Adminer database client at `http://localhost:8080/`
- the page hit service at `http://localhost:8001/`
- a test page web server at `http://localhost:8888/`

Different ports can be defined in the `.env` file if you have clashes.

Visit http://localhost:8888/page1.html or http://localhost:8888/page2.html to view page counters. Refresh and watch the counter increase.

You can examine the database data using the Adminer panels at http://localhost:8080/. Log on with the credentials specified in `.env`:

- System: **MySQL**
- Server: **host.docker.internal** (or your network IP address if not using Docker Desktop)
- Username: **pagehituser**
- Password: **pagehitpw**
- Database: **pagehitorm**

Then click **select** next to the `hits` table.

Sequelize ORM Functionality

The `lib/pagehit.js` file handles all Sequelize functionality. It loads the required modules and extracts the configuration parameters from the `.env` file using the `dotenv` [32] module:

```
import dotenv from 'dotenv';
```

[32.] https://www.npmjs.com/package/dotenv

```
import Sequelize from 'sequelize';
import httpReferrer from './httpreferrer.js';

// load .env configuration
dotenv.config();
```

(There's no need to *import* the *mysql2* module, as *Sequelize* loads it.)

The database name, user, and password connection parameters are passed to the *Sequelize* object constructor. A fourth options object defines the database type, host, and port:

```
// initialize ORM connection
const sequelize = new Sequelize(
  process.env.MYSQL_DATABASE,
  process.env.MYSQL_USER,
  process.env.MYSQL_PASSWORD,
  {
    host: process.env.MYSQL_HOST,
    port: process.env.MYSQL_PORT,
    dialect: 'mysql'
  }
);
```

Rather than defining a table, you create a JavaScript class from a *Sequelize.Model* [33] class. The static *init()* method[34] defines the property data types and indexes (note that each model has a default *id* , *createdAt* , and *updatedAt* date):

```
// define Hit class
class Hit extends Sequelize.Model {}
Hit.init(
  {

    hash: {
      type: Sequelize.STRING(32),
```

[33.] https://sequelize.org/master/manual/model-basics.html
[34.] https://sequelize.org/master/class/lib/model.js~Model.html#static-method-init

```
      allowNull: false
    },
    ip: {
      type: Sequelize.STRING(15),
      allowNull: true
    },
    ua: {
      type: Sequelize.STRING(200),
      allowNull: true
    }

  },
  {
    indexes: [
      { fields: [ 'hash', 'createdAt' ] }
    ],
    sequelize,
    modelName: 'hit'
  }

);
```

The asynchronous *sync()* method[35] synchronizes all data models with the database. In this case, a *hits* table is defined from the *Hit* model:

```
// synchronize model with database
await sequelize.sync();
```

Like before, *lib/pagehit.js* exports a single default asynchronous function. It generates a *hash* from the referring page's URL, returns when no referrer is found, and determines the browser's IP address (*ip*), and user agent (*ua*):

```
// count handler
export default async function(req) {

  // hash of referring URL
  const hash = httpReferrer(req);
```

---

[35]. https://sequelize.org/master/class/lib/sequelize.js~Sequelize.html#instance-method-sync

```
  // no referrer?
  if (!hash) return null;

  // fetch browser IP address and user agent
  const
    ipRe  = req.ip.match(/(?:\d{1,3}\.){3}\d{1,3}/),
    ip    = ipRe?.[0] || null,
    ua    = req.get('User-Agent') || null;
```

A new `Hit` record is created[36] with the data:

```
try {

  // store page hit
  await Hit.create(
    { hash, ip, ua }
  );
```

A count of all items with the same `hash` is then returned:

```
  // fetch page hit count
  const res = await Hit.findAndCountAll({
    where: { hash }
  });

  return res?.count;
```

An error is thrown if any operation fails:

```
  }
  catch (err) {
    throw new Error('DB error', { cause: err });
  }

}
```

As before, the main `index.js` script loads the `lib/pagehit.js` module, sets

36. https://sequelize.org/master/class/lib/model.js~Model.html#static-method-create

`req.count` in a middleware function, and outputs it in a generated SVG in the `/hit.svg` route.

How to Choose the Right Database

An SQL database such as MySQL is the best option when requirements are clearly defined and data integrity is essential—such as for banking, ecommerce, stock control, and so on. A money transfer requires an amount to be debited from one account and credited to another: transactions guarantee that both or neither update is successful.

A NoSQL database such as MongoDB could be ideal for projects where organic data flexibility is important—such as content management, social networks, web analytics, and so on.

In general:

- A NoSQL database can be easier to use at the start of a project, but may become more difficult as you identify data relationships.
- An SQL database requires more careful data planning up front, but this can return dividends toward the end of a project—*(unless requirements change radically!)*

Complex projects *could* benefit from using two or more databases. For example, a blog stored in MySQL could use Elasticsearch for Google-like search queries. However, maintaining data integrity between two or more databases is complex and cumbersome. It may also be unnecessary, because the distinction between SQL and NoSQL has blurred:

- some SQL databases have adopted NoSQL features, such as JSON and XML fields
- some NoSQL databases have adopted SQL features, such as JOINs and transactions

Research the options, browse usage reports[37], and consult others to make sure a database has the features and support you need. Try to abstract your data manipulation code so it becomes easier to switch to another system if that becomes necessary.

Native vs ORM Drivers

Think of an ORM as an abstract database framework. The benefits include:

- They can be easier to learn than specific SQL or NoSQL dialects.
- Development time is reduced, because a good ORM will manage security and data integrity.
- You can create data models in the application code. There's no need to directly create or alter tables.
- ORMs track changes and can migrate schemas as necessary.
- ORMs support multiple databases, which could be important if you're distributing web software for others to install.

The downsides of an ORM:

- They can still be difficult to learn. The Sequelize manual[38] is daunting, and that knowledge won't be applicable elsewhere.
- An ORM won't save you from poor data decisions.
- More complex queries can be difficult to express.
- ORMs are slower, and queries won't necessarily be optimized.
- You'll be unable to use advanced options provided in a specific database.
- Database-related bugs may be more difficult to debug.

An ORM can be ideal for prototypes and smaller projects. Native drivers with optimized, hand-crafted queries are better for larger projects where data requirements are more critical.

If I could offer one piece of advice: *learn SQL.*

Unlike most development technologies, SQL has persisted for half a century,

[37.] https://db-engines.com/en/ranking
[38.] https://sequelize.org/master/

and the skills are transferable to other databases. You'll become more adept at modeling data and creating efficient applications that require less code. You'll have fewer reasons to consider an ORM.

Exercises

Adapt any of the page hit counter projects so different routes can return:

- Page hits during the past 24 hours.
- Page hits from the current IP address.
- Page hits from the same browser (Chrome, Edge, Safari, Firefox, etc.). This is tricky! Browser user agent strings purposely obfuscate the application! You may also need to parse and output another field to make searches more efficient.

Summary

Databases are a core web application technology. If your database functions well, it won't necessarily matter whether you write the business logic using Node.js, PHP, Python, Ruby, or .NET.

You need to start somewhere, so pick a database and create example projects. Choosing an SQL database with a native driver will have a steep learning curve, but persevere! *It will make you a better web developer*.

Quiz

1. SQL is short for:

- a. Simple Query Language
- b. Structured Query Language
- c. Statistical Query Language
- d. Small & Quick Language

2. A MySQL database table can store:

a. table data
b. JSON data
c. XML data
d. all of the above

3. A MongoDB database can:

a. be used without a data schema
b. be used with a data schema
c. join data in two collections
d. all of the above

4. ORM is short for:

a. object-relational mapping
b. object-reference model
c. ordered-reference map
d. ordered-results management

5. A database index:

a. defines data in a specific order
b. is automatically used when required
c. should make queries faster
d. all of the above

Using WebSockets

This chapter demonstrates how to use **WebSockets**—a technology that makes it possible to open a two-way, interactive communication channel between a browser and a server. In the past, this was difficult to achieve on most platforms, and often required a third-party service. Node.js makes it easy, although we'll delve into some deeper challenges.

 Skip Ahead?

> It's possible to become a respected senior developer without touching WebSockets! You can skip this chapter, but the technology opens a world of opportunities you may not have considered before.

What Are WebSockets?

The web is a request–response communication platform. Your browser requests a web page and receives HTML as the response from a web server. The page may reference assets such as images, fonts, CSS, and JavaScript; the browser makes an additional request for each.

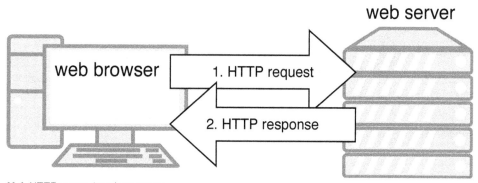

11-1. HTTP request and response

The browser initiates every request. *A web server can't arbitrarily push data to a user.* It must be requested first.

Ajax techniques can be used to make web apps look as though they update in real time by initiating a polling request every few seconds. This can check for new data from a web server and update the DOM as necessary.

Few apps need go beyond this request–response model, because data changes infrequently in a typical web application. However, it's not ideal for true real-time applications such as stock price dashboards, chat apps, and multiplayer games. Polling every second would be inefficient at certain times, and too slow at others. It's also difficult for a server to determine what changed between two polling intervals: every browser could be asking for different data.

WebSockets[1] provide a solution for real-time apps. The browser makes an initial WebSocket request, which opens a communication channel. At that point, either the browser or server can send a message that raises an event on the other device.

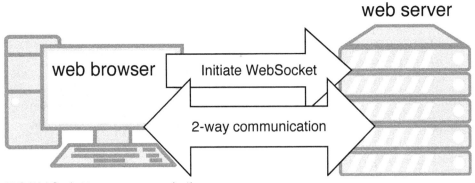

11-2. WebSocket two-way communication

Two things to be aware of:

- A browser can only send a message to the WebSocket server.
- The WebSocket server can send a message to any of its connected clients.

One browser can't directly message another. It can only send a message to the central WebSocket server and hope it gets forwarded as necessary.

---

[1] https://developer.mozilla.org/Web/API/WebSockets_API

Example WebSocket Chat Application

The sections below explain how to create a simple real-time chat app using WebSockets. Chat apps are the "Hello, World!" of WebSocket demonstrations, so I apologize for being unoriginal—but they show the concepts without too much code.

To get started, navigate to the *wschat* code directory[2] in your terminal and install the Node.js dependencies with `npm install`.

Run the application with `npm start`.

Open http://localhost:3000/ in a number of browser tabs (you can also define your chat name on the query string—such as http://localhost:3000/?Craig). Type something in one window, then press **SEND** or hit `Enter`, and you'll see it appear in every window.

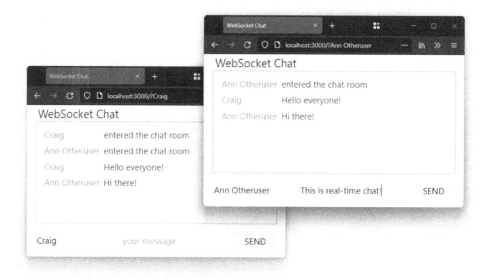

11-3. Chat windows

View the video[3] to see the chat application in action.

2. https://github.com/spbooks/ultimatenode1/tree/main/ch11/wschat

WebSocket Walkthrough

The application works by starting two server processes in the *index.js* file:

- An Express app with an EJS template to serve a single page with client-side HTML, CSS, and JavaScript. This runs at **http://localhost:3000/** and uses the browser WebSocket API[4] to send and receive messages.

- A WebSocket server, which listens for incoming client connections, receives messages, sends messages, and monitors disconnections. This runs at **ws://localhost:3001/** and uses theNode.js *ws* library[5].

  - A *"connection"* event is raised when a connection is received from a browser. The handler function receives a *socket* object used to communicate with that individual device.

  - A *socket* *"message"* event is raised when a client sends a message. The chat application's handler function broadcasts that message to every connected client.

  - A *socket* *"close"* event is raised when the client disconnects (typically when the browser tab is closed or refreshed).

Here's the full **server JavaScript code**:

```
// WebSocket server
import WebSocket, { WebSocketServer } from 'ws';

const ws = new WebSocketServer({ port: cfg.wsPort });

// client connection
ws.on('connection', (socket, req) => {
```

---

3. https://spnt.co/nodevid17
4. https://developer.mozilla.org/Web/API/WebSocket
5. https://www.npmjs.com/package/ws

```
    console.log(`connection from ${ req.socket.remoteAddress }`);

    // received message
    socket.on('message', (msg, binary) => {

      // broadcast to all clients
      ws.clients.forEach(client => {
        client.readyState === WebSocket.OPEN && client.send(msg, { binary });
      });

    });

    // closed
    socket.on('close', () => {
      console.log(`disconnection from ${ req.socket.remoteAddress }`);
    });

  });
```

The **client-side browser** JavaScript:

1   caches HTML *dom* nodes for later use

2   sets a default username from the query string or a random string

3   determines the *ws://* WebSocket connection address using the page's

domain plus the port defined in the HTML page template

```
// get page DOM nodes
const dom = { form: 0, chat: 0, name: 0, message: 0 };
for (let n in dom) dom[n] = document.getElementById(n);

// set user's name
dom.name.value = decodeURIComponent(location.search.trim().slice(1,1 + window.
↪cfg.nameLen)) || 'Anonymous' + Math.floor(Math.random() * 99999);

wsInit(`ws://${ location.hostname }:${ window.cfg.wsPort }`);
```

A *wsInit()* function is called with the WebSocket server address to initiate
the connection. An *open* event is triggered when a connection is established.

At this point, the handler function sends an "entered the chat room" message
by calling *sendMessage()* :

```
// handle WebSocket communication
function wsInit(wsServer) {

  const ws = new WebSocket(wsServer);

  // connect to server
  ws.addEventListener('open', () => {
    sendMessage('entered the chat room');
  });
```

The *sendMessage()* function fetches the user's name and message from the
HTML form, although the message can be overridden by any passed *setMsg*
argument. The values are converted to a JSON object that's sent over the
WebSocket channel using its *ws.send()* method:

```
// send message
function sendMessage(setMsg) {

  let
    name = dom.name.value.trim(),
    msg =  setMsg || dom.message.value.trim();

  name && msg && ws.send( JSON.stringify({ name, msg }) );

}
```

The message is received by the server's *"message"* handler and broadcast to
every connected client—*including the client that sent the message.* This
triggers a *"message"* event on each client, with the event's *data* property set
to the original JSON. The handler function parses this back to a JavaScript
object and updates the chat window:

```
// receive message
ws.addEventListener('message', e => {
```

```
  try {

    const
      chat = JSON.parse(e.data),
      name = document.createElement('div'),
      msg  = document.createElement('div');

    name.className = 'name';
    name.textContent = (chat.name || 'unknown');
    dom.chat.appendChild(name);

    msg.className = 'msg';
    msg.textContent = (chat.msg || 'said nothing');
    dom.chat.appendChild(msg).scrollIntoView({ behavior: 'smooth' });

  }
  catch(err) {
    console.log('invalid JSON', err);
  }

});
```

Finally, new messages are sent using *sendMessage()* whenever the form's
"submit" handler is triggered:

```
// form submit
dom.form.addEventListener('submit', e => {
  e.preventDefault();
  sendMessage();
  dom.message.value = '';
  dom.message.focus();
}, false);
```

This chapter's second video[6] also explains the basics of the chat application's
functionality.

Advanced WebSockets Considerations

WebSocket technology is easy in Node.js: one device sends a message using a

---

6. https://spnt.co/nodevid18

`send()` method, which triggers a `"message"` event on the other. How each device creates and responds to messages is more challenging.

Consider an online multiplayer game. The game could have many *universes* playing separate instances of the game—such as `universeA`, `universeB`, and `universeC`. Each player can connect to a single universe:

- `universeA` : joined by `player1`, `player2`, and `player3`
- `universeB` : joined by `player99`

You could do one of the following:

1. **Use a separate WebSocket server for each game universe.**

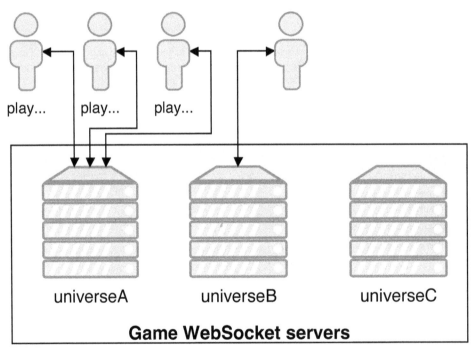

11-4. Using multiple WebSocket servers

This would make user management easy: a player action in `universeA` would never be seen by those in `universeB`. However, launching and managing separate server instances could be difficult. Would you stop `universeC` because it has no players, or continue to pay for that resource?

2   **Use a single WebSocket server for all game universes.**

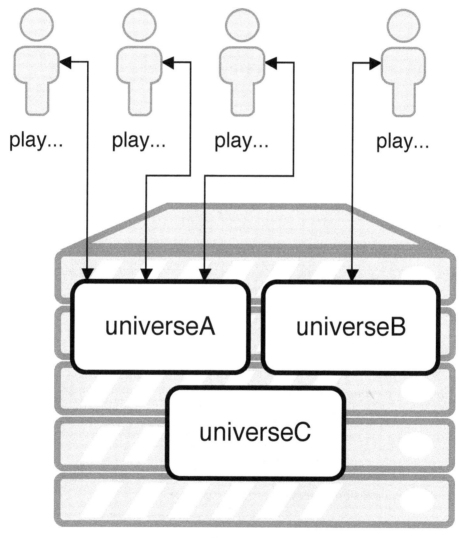

11-5. Using a single WebSocket server

This would use fewer resources and be easier to manage, but the WebSocket server must record which universe each player joins. When `player1` performs an action, it should only be broadcast to `player2` and `player3` —not `player99`.

You must then consider game mechanics and messaging efficiency. For example:

- How do you synchronize a player's actions across all client devices?

- If `player1` can't currently be seen by `player2` (because they're in another room), is it necessary for `player2` to receive a message about their actions?

- How do you cope with network latency—*or communication lag?* Would someone with a faster machine and connection have an unfair advantage?

Fast action games have to make compromises. In essence, you're playing the game on your local device but some objects are *influenced* by the activities of other people. Rather than sending the exact position of every object at all times, games can send simpler, less frequent messages. For example:

- `objectX` has appeared at pointX
- `objectY` has a new direction and velocity
- `objectZ` has been destroyed

… and so on.

Each client game fills in the gaps. When `objectZ` explodes, it rarely matters whether the explosion looks the same on every device.

This all explains why you were unfairly beaten in your favorite game by a seemingly invisible player!

Multiple WebSocket Servers

The example chat application can cope with dozens of concurrent users, but at some point, it'll crash as popularity rises. More RAM can help, but there are limits. *You'll eventually require another server.*

Each WebSocket server can only manage its own connected clients. A message sent from a user to *serverA* wouldn't be broadcast to those connected to *serverB* . It may be necessary to implement backend, pub–sub messaging systems such as Kafka, Redis, or RabbitMQ.

 What is Pub–sub?

> Publisher-subscriber services provide asynchronous communication services. An application can send (*publish*) a message to the pub–sub system. Applications can *subscribe* to those messages and be instantly alerted when a new one arrives.

Therefore:

1. WebSocket *serverA* wants to send a message to all clients. It publishes the message on the pub–sub system.

2. All WebSocket servers subscribed to the pub–sub system receive a new message event (including *serverA*). Each can handle the message and broadcast it to their connected clients as necessary.

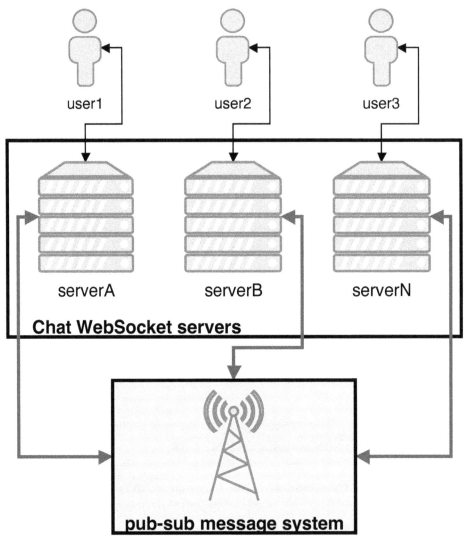

11-6. Multiple WebSocket servers with a pub–sub system

The example real-time quiz at the end of this course uses PostgreSQL to provide pub–sub functionality.

Exercise

Adapt the chat application to store the most recent 30 messages and send them to any new user entering the chat room. For big bonus points, store message data in a database so it persists between application restarts.

You could also experiment with sending different types of messages. For example, allow private messages to be sent to a single user.

Summary

Node.js makes it easy to handle WebSockets. It won't make real-time applications easier to design or code, but the technology won't hold you back.

Note that *ws* isn't the only Node.js option. Almost 1,000 other WebSocket packages are available. Some provide their own browser client libraries or integrate with JavaScript frameworks to make usage easier.

You could also consider server-sent events[7] if your app only needs to receive updates from a central server.

Quiz

1. WebSockets offer:

    a. two-way browser/server communication
    b. real-time messaging
    c. an event-driven API
    d. all of the above

2. A WebSocket server:

    a. responds to connections and messages from clients
    b. passes connection requests to and from a web server
    c. initiates the WebSocket connection
    d. all of the above

[7.] https://developer.mozilla.org/Web/API/Server-sent_events/Using_server-sent_events

3. A message sent on a WebSocket connection must be:

   a. a string
   b. JSON
   c. binary data
   d. any text or binary data

4. A browser WebSocket client can message another user's browser by:

   a. sending a direct peer-to-peer message that bypasses the server
   b. sending a message to the WebSocket server that forwards as necessary
   c. adding the other user's IP address to the message
   d. all of the above

5. Which best describes WebSocket code as used in the `ws` library?

   a. a `message()` call that triggers a `"sent"` event on the other device
   b. a `send()` call that triggers a `"message"` event on the other device
   c. a `transmit()` call that triggers a `"received"` event on the other device
   d. a `send()` call that triggers a `"receive"` event on the other device

Chapter

Useful
Node.js APIs

12

This chapter demonstrates a selection of regularly used APIs that are built in to the standard Node.js runtime. You've seen some in previous chapters of this book, but I hope the following sections will pique your interest and encourage you to browse the Node.js API documentation.

 Module node: URL Imports

Node.js 14 and above support *node:* imports[1] for both ESM and CommonJS modules. Rather than using the API's module name:

```
import path from 'path';
```

... you can reference it using an absolute *node:* URL:

```
import path from 'node:path';
```

This might be practical if you had another module named *path* or want to distinguish built-in APIs in your code.

Process

The *process* object[2] provides information about your Node.js application as well as control methods. *process* is available globally: you can use it without *import* , although the Node.js documentation recommends you explicitly reference it:

```
import process from 'process';
```

We've used *process.argv* in other scripts to fetch commmand-line arguments:

```
const firstArg = process.argv[2];
```

---

1. https://nodejs.org/api/esm.html#node-imports
2. https://nodejs.org/dist/latest/docs/api/process.html

process.argv returns an array where the first two items are the Node.js executable path and the script name. The item at index 2 is the first argument passed.

Other useful properties and methods include:

- *process.env* : returns an object containing environment name/value pairs—such as *process.env.NODE_ENV* .

- *process.cwd()* : returns the current working directory.

- *process.platform* : returns a string identifying the operating system: *'aix'* , *'darwin'* (macOS), *'freebsd'* , *'linux'* , *'openbsd'* , *'sunos'* , or *'win32'* (Windows).

- *process.uptime()* : returns the number of seconds the Node.js process has been running.

- *process.cpuUsage()* : returns the user and system CPU time usage of the current process—such as *{ user: 12345, system: 9876 }* . Pass the object back to the method to get a relative reading.

- *process.memoryUsage()* : returns an object describing memory usage in bytes.

- *process.version* : returns the Node.js version string—such as *18.0.0* .

- *process.report* : generates a diagnostic report.

- *process.exit(code)* : exits the current application. Use an exit code of *0* to indicate success or an appropriate error code[3] where necessary.

process is also an event emitter (see the "Events" section): you can attach event handler functions to events such as *'beforeExit'* to clean up before

---

3. https://nodejs.org/dist/latest/docs/api/process.html#exit-codes

the process terminates. For example:

```
// clean up when the Node.js process terminates
process.on('beforeExit', code => {
  // ...
});
```

 exit Events

> You can also define an *exit* handler function. However, this can't
> run asynchronous functions such as disconnecting from a database
> or outputting a log file, because the Node.js event loop will end on
> the current iteration and the program will terminate.

OS

The *os* API[4] has similarities to *process* (see the "Process" section above),
but it can also return the following:

- *os.cpus()* : returns an array of objects with information about each logical
  CPU core. The "Clusters" section below references *os.cpus()* to fork the
  process. On a 16-core CPU, you'd have 16 instances of your Node.js
  application running to improve performance.

- *os.hostname()* : the OS host name.

- *os.version()* : a string identifying the OS kernel version.

- *os.homedir()* : the full path of the user's home directory.

- *os.tmpdir()* : the full path of the operating system's default temporary file
  directory.

---

4. https://nodejs.org/dist/latest/docs/api/os.html

- `os.uptime()` : the number of seconds the OS has been running.

Util

The `util` module provides an assortment of useful JavaScript methods. One of the most useful is `util.promisify(function)` , which takes an error-first callback style function and returns a promise-based function. (See the code in Chapter 9 for a demonstration[5].)

Further methods include:

- `util.callbackify(function)` : takes a function that returns a promise and returns a callback-based function.

- `util.isDeepStrictEqual(object1, object2)` : returns `true` when there's a deep equality between two objects (all child properties must match).

- code>util.format(format, [args]): returns a string using a printf-like format.

- `util.inspect(object, options)` : returns a string representation of an object for debugging. This is similar to using `console.dir(object, { depth: null, color: true });` .

- `util.stripVTControlCharacters(str)` : strips ANSI escape codes from a string.

`util.types` > provides type checking for common JavaScript and Node.js values. For example:

```
import util from 'util';

util.types.isDate( new Date() ); // true
util.types.isMap( new Map() );   // true
util.types.isRegExp( /abc/ ); // true
```

[5.] https://github.com/spbooks/ultimatenode1/blob/main/ch09/05-promisify.mjs

```
util.types.isAsyncFunction( async () => {} ); // true
```

URL

URL [6] is another global object that lets you safely create, parse, and modify web URLs. For example:

```
const myURL = new URL('https://example.org:8000/path/?abc=123#target');
console.dir(myURL, { depth: null, color: true });
```

The code above outputs this:

```
{
  href: 'https://example.org:8000/path/?abc=123#target',
  origin: 'https://example.org:8000',
  protocol: 'https:',
  username: '',
  password: '',
  host: 'example.org:8000',
  hostname: 'example.org',
  port: '8000',
  pathname: '/path/',
  search: '?abc=123',
  searchParams: URLSearchParams { 'abc' => '123' },
  hash: '#target'
}
```

You can view and change any property. For example:

```
myURL.port = 8001;
console.log( myURL.href );
// https://example.org:8001/path/?abc=123#target
```

You can then use the *URLSearchParams* API[7] to modify query string values. For example:

---

6. https://nodejs.org/dist/latest/docs/api/url.html
7. https://nodejs.org/dist/latest/docs/api/url.html#class-urlsearchparams

```
myURL.searchParams.delete('abc');
myURL.searchParams.append('xyz', 987);
console.log( myURL.search );
// ?xyz=987
```

There are also methods for converting file system paths to URLs[8] and back again[9].

The *dns* module[10] provides name resolution functions so you can look up the IP address, name server, TXT records, and other domain information.

File System

The *fs* API[11] can create, read, update, and delete files, directories, and permissions. Recent releases of the Node.js runtime provide promise-based functions in *fs/promises* [12], which make it easier to manage asynchronous file operations.

 fs and path

You'll often use *fs* in conjunction with *path* [13] to resolve file names on different operating systems.

The example code has a *filecompress* project[14], which compresses a text file (typically HTML, CSS, or JS) by removing whitespace and comments. *(It's a demonstration—so please don't use it on real files! The compression process is simplistic and will mangle some files.)*

---

8. https://nodejs.org/dist/latest/docs/api/url.html#urlpathtofileurlpath
9. https://nodejs.org/dist/latest/docs/api/url.html#urlfileurltopathurl
10. https://nodejs.org/dist/latest/docs/api/dns.html
11. https://nodejs.org/dist/latest/docs/api/fs.html
12. https://nodejs.org/dist/latest/docs/api/fs.html#promises-api
13. https://nodejs.org/dist/latest/docs/api/path.html
14. https://github.com/spbooks/ultimatenode1/tree/main/ch12/filecompress

The project has a *lib/fileinfo.js* module that returns information about a
file system object using the *stat* and *access* methods:

```
// fetch file information
import { constants as fsConstants } from 'fs';
import { access, stat } from 'fs/promises';

export async function getFileInfo(file) {

  const fileInfo = {};

  try {
    const info = await stat(file);
    fileInfo.isFile = info.isFile();
    fileInfo.isDir = info.isDirectory();
  }
  catch (e) {
    return { new: true };
  }

  try {
    await access(file, fsConstants.R_OK);
    fileInfo.canRead = true;
  }
  catch (e) {}

  try {
    await access(file, fsConstants.W_OK);
    fileInfo.canWrite = true;
  }
  catch (e) {}

  return fileInfo;

}
```

When passed a filename, the function returns an object with information
about that file. For example:

```
{
  isFile: true,
```

```
    isDir: false,
    canRead: true,
    canWrite: true
}
```

The main *filecompress.js* script uses *path.resolve()* to resolve input and output filenames passed on the command line into absolute file paths, then fetches information using *getFileInfo()* above:

```
#!/usr/bin/env node
import path from 'path';
import { readFile, writeFile } from 'fs/promises';
import { getFileInfo } from './lib/fileinfo.js';

// check files
let
  input = path.resolve(process.argv[2] || ''),
  output = path.resolve(process.argv[3] || ''),
  [ inputInfo, outputInfo ] = await Promise.all([ getFileInfo(input),
  ↪getFileInfo(output) ]),
  error = [];
```

The code validates the paths and terminates with error messages if necessary:

```
// use input file name when output is a directory
if (outputInfo.isDir && outputInfo.canWrite && inputInfo.isFile) {
  output = path.resolve(output, path.basename(input));
}

// check for errors
if (!inputInfo.isFile || !inputInfo.canRead) error.push(`cannot read input file
↪${ input }`);
if (input === output) error.push('input and output files cannot be the same');

if (error.length) {

  console.log('Usage: ./filecompress.js [input file] [output file|dir]');
  console.error('\n  ' + error.join('\n  '));
  process.exit(1);
```

```
}
```

The whole file is then read into a string named *content* using *readFile()* [15]:

```
// read file
console.log(`processing ${ input }`);
let content;

try {
  content = await readFile(input, { encoding: 'utf8' });
}
catch (e) {
  console.log(e);
  process.exit(1);
}

let lengthOrig = content.length;
console.log(`file size  ${ lengthOrig }`);
```

JavaScript regular expressions then remove comments and whitespace:

```
// compress content
content = content
  .replace(/\n\s+/g, '\n')                 // trim leading space from lines
  .replace(/\/\/.*?\n/g, '')               // remove inline // comments
  .replace(/\s+/g, ' ')                    // remove whitespace
  .replace(/\/\*.*?\*\//g, '')             // remove /* comments */
  .replace(/<!--.*?-->/g, '')              // remove <!-- comments -->
  .replace(/\s*([<>(){}}[\]])\s*/g, '$1')  // remove space around brackets
  .trim();

let lengthNew = content.length;
```

The resulting string is output to a file using *writeFile()* , and a status message shows the saving:

```
let lengthNew = content.length;
```

[15.] https://nodejs.org/dist/latest/docs/api/fs.html#fspromisesreadfilepath-option

```
// write file
console.log(`outputting ${output}`);
console.log(`file size  ${ lengthNew } - saved ${ Math.round((lengthOrig -
↪lengthNew) / lengthOrig * 100) }%`);

try {
  content = await writeFile(output, content);
}
catch (e) {
  console.log(e);
  process.exit(1);
}
```

Run the project code with an example HTML file:

```
node filecompress.js ./test/example.html ./test/output.html
```

12-1. filecompress.js output

View the demonstration video[16] to see the code in action.

Events

You often need to execute multiple functions when something occurs. For example, a user registers on your app, so the code must add their details to a database, start a new logged-in session, and send a welcome email:

```
// example pseudo code
async function userRegister(name, email, password) {
```

```
  try {

    await dbAddUser(name, email, password);
    await new UserSession(email);
    await emailRegister(name, email);

  }
  catch (e) {
    // handle error
  }

}
```

This series of function calls is tightly coupled to user registration. Further activities incur further function calls. For example:

```
// updated pseudo code
try {

  await dbAddUser(name, email, password);
  await new UserSession(email);
  await emailRegister(name, email);

  await crmRegister(name, email); // register on customer system
  await emailSales(name, email);  // alert sales team

}
```

You could have dozens of calls managed in this single, ever-growing code block.

The Node.js Events API[17] provides an alternative way to structure the code using a publish–subscribe pattern. The *userRegister()* function can *emit* an event—perhaps named *newuser* —after the user's database record is created.

Any number of event handler functions can subscribe and react to *newuser* events; there's no need to change the *userRegister()* function. Each handler runs independently of the others, so they could execute in any order.

[17.] https://nodejs.org/dist/latest/docs/api/events.html

Events in Client-side JavaScript

Events and handler functions are frequently used in client-side JavaScript—for example, to run a function when the user clicks an element:

```
// client-side JS click handler
  document.getElementById('myelement').addEventListener('click', e => {

    // output information about the event
    console.dir(e);

  });
```

In most situations, you're attaching handlers for user or browser events, although you can raise your own custom events[18]. Event handling in Node.js is conceptually similar, but the API is different.

Objects that emit events must be instances of the Node.js *EventEmitter* class. These have an *emit()* method to raise new events and an *on()* method for attaching handlers.

The event example project[19] provides a class that triggers a *tick* event on predefined intervals. The *./lib/ticker.js* module exports a *default class* that *extends EventEmitter*:

```
// emits a 'tick' event every interval
import EventEmitter from 'events';
import { setInterval, clearInterval } from 'timers';

export default class extends EventEmitter {
```

Its *constructor* must call the parent constructor. It then passes the *delay* argument to a *start()* method:

18. https://developer.mozilla.org/docs/Web/API/CustomEvent
19. https://github.com/spbooks/ultimatenode1/tree/main/ch12/event

```
constructor(delay) {
  super();
  this.start(delay);
}
```

The *start()* method checks delay is valid, resets the current timer if necessary, and sets the new *delay* property:

```
start(delay) {

  if (!delay || delay == this.delay) return;

  if (this.interval) {
    clearInterval(this.interval);
  }

  this.delay = delay;
```

It then starts a new interval timer that runs the *emit()* method with the event name *"tick"* . Subscribers to this event receive an object with the delay value and number of seconds since the Node.js application started[20]:

```
// start timer
this.interval = setInterval(() => {

  // raise event
  this.emit('tick', {
    delay:  this.delay,
    time:   performance.now()
  });

}, this.delay);

  }

}
```

The main *event.js* entry script imports the module and sets a *delay* period

---

[20]. https://nodejs.org/dist/latest/docs/api/perf_hooks.html#performancenow

of one second (`1000` milliseconds):

```
// create a ticker
import Ticker from './lib/ticker.js';

// trigger a new event every second
const ticker = new Ticker(1000);
```

It attaches handler functions triggered every time a `tick` event occurs:

```
// add handler
ticker.on('tick', e => {
  console.log('handler 1 tick!', e);
});

// add handler
ticker.on('tick', e => {
  console.log('handler 2 tick!', e);
});
```

A third handler triggers on the first `tick` event only using the `once()` method:

```
// add handler
ticker.once('tick', e => {
  console.log('handler 3 tick!', e);
});
```

Finally, the current number of listeners is output:

```
// show number of listeners
console.log(`listeners: ${ ticker.listenerCount('tick') }`);
```

Run the project code with `node event.js` .

The output shows handler 3 triggering once, while handler 1 and 2 run on every `tick` until the app is terminated.

```
craig@craigdev: ~/apps/event    ×    +  ∨                —   □   ×

craig@craigdev:~/apps/event$ node event.js
listeners: 3
handler 1 tick! { delay: 1000, time: 1046.2753809999995 }
handler 2 tick! { delay: 1000, time: 1046.2753809999995 }
handler 3 tick! { delay: 1000, time: 1046.2753809999995 }
handler 1 tick! { delay: 1000, time: 2047.6881189999986 }
handler 2 tick! { delay: 1000, time: 2047.6881189999986 }
handler 1 tick! { delay: 1000, time: 3048.5061679999926 }
handler 2 tick! { delay: 1000, time: 3048.5061679999926 }
^C
craig@craigdev:~/apps/event$
```

12-2. An event example

Press `Ctrl`|`Cmd` + `C` to terminate the application.

View the demonstration video[21] to see the code in action.

Streams

The file system example code above (in the "File System" section) reads a whole file into memory before outputting the minified result. What if the file was larger than the RAM available? The Node.js application would fail with an "out of memory" error.

The solution is **streaming**. This processes incoming data in smaller, more manageable chunks. A stream can be:

- **readable**: from a file, a HTTP request, a TCP socket, stdin, etc.
- **writable**: to a file, a HTTP response, TCP socket, stdout, etc.
- **duplex**: a stream that's both readable and writable
- **transform**: a duplex stream that transforms data

Each chunk of data is returned as a *Buffer* object[22], which represents a

---

[21.] https://spnt.co/nodevid20

fixed-length sequence of bytes. You may need to convert this to a string or another appropriate type for processing.

The example code has a *filestream* project[23] which uses a transform stream to address the file size problem in the *filecompress* project. As before, it accepts and validates *input* and *output* filenames before declaring a *Compress* class, which extends *Transform* :

```
import { createReadStream, createWriteStream } from 'fs';
import { Transform } from 'stream';

// compression Transform
class Compress extends Transform {

  constructor(opts) {
    super(opts);
    this.chunks = 0;
    this.lengthOrig = 0;
    this.lengthNew = 0;
  }

  _transform(chunk, encoding, callback) {

    const
      data = chunk.toString(),               // buffer to string
      content = data
        .replace(/\n\s+/g, '\n')             // trim leading spaces
        .replace(/\/\/.*?\n/g, '')           // remove // comments
        .replace(/\s+/g, ' ')                // remove whitespace
        .replace(/\/\*.*?\*\//g, '')         // remove /* comments */
        .replace(/<!--.*?-->/g, '')          // remove <!-- comments -->
        .replace(/\s*([<>(){}}[\]])\s*/g, '$1') // remove bracket spaces
        .trim();

    this.chunks++;
    this.lengthOrig += data.length;
    this.lengthNew += content.length;
```

22. https://nodejs.org/dist/latest/docs/api/buffer.html
23. https://github.com/spbooks/ultimatenode1/tree/main/ch12/filestream

```
    this.push( content );
    callback();

  }

}
```

The `_transform` method is called when a new *chunk* of data is ready. It's received as a *Buffer* object that's converted to a string, minified, and output using the `push()` method. A `callback()` function is called once chunk processing is complete.

The application initiates file read and write streams and instantiates a new *compress* object:

```
// process stream
const
  readStream = createReadStream(input),
  writeStream = createWriteStream(output),
  compress = new Compress();

console.log(`processing ${ input }`);
```

The incoming file read stream has `.pipe()` methods defined, which feed the incoming data through a series of functions that may (or may not) alter the contents. The data is *piped* through the *compress* transform before that output is *piped* to the writeable file. A final `on('finish')` event handler function executes once the stream has ended:

```
readStream.pipe(compress).pipe(writeStream).on('finish', () => {

  console.log(`file size  ${ compress.lengthOrig }`);
  console.log(`output     ${ output }`);
  console.log(`chunks     ${ compress.chunks }`);
  console.log(`file size  ${ compress.lengthNew } - saved ${ Math.round((
  ↪compress.lengthOrig - compress.lengthNew) / compress.lengthOrig * 100) }%`);

});
```

Run the project code with an example HTML file of any size:

```
node filestream.js ./test/example.html ./test/output.html
```

```
craig@craigdev:~/apps/filestream$ node filestream.js ./test/example.html ./test/output.html
processing /home/craig/apps/filestream/test/example.html
file size  760802
output     /home/craig/apps/filestream/test/output.html
chunks     12
file size  598267 - saved 21%
craig@craigdev:~/apps/filestream$
```

12-3. filestream.js output

View the demonstration video[24] to see the code in action.

This is a small demonstration of Node.js streams. Stream handling is a complex topic, and you may not use them often. In some cases, a module such as Express uses streaming under the hood but abstracts the complexity from you.

You should also be aware of data chunking challenges. A chunk could be any size and split the incoming data in inconvenient ways. Consider minifying this code:

```
<script type="module">
  // example script
  console.log('loaded');
</script>
```

Two chunks could arrive in sequence:

```
<script type="module">
  // example
```

And:

```
script
  console.log('loaded');
</script>
```

Processing each chunk independently results in the following invalid minified script:

```
<script type="module">script console.log('loaded');</script>
```

The solution is to pre-parse each chunk and split it into whole sections that can be processed. In some cases, chunks (or parts of chunks) will be added to the start of the next chunk.

Minification is best applied to whole lines, although an extra complication occurs because `<!-- -->` and `/* */` comments can span more than one line. Here's a possible algorithm for each incoming chunk:

1. Append any data saved from the previous chunk to the start of the new chunk.

2. Remove any whole `<!--` to `-->` and `/*` to `*/` sections from the chunk.

3. Split the remaining chunk into two parts, where *part2* starts with the first `<!--` or `/*` found. If either exists, remove further content from *part2* except for that symbol.

If neither is found, split at the last carriage return character. If none is found, set *part1* to an empty string and *part2* to the whole chunk.

If *part2* becomes significantly large—perhaps more than 100,000 characters because there are no carriage returns—append *part2* to *part1* and set *part2* to an empty string. This will ensure saved parts can't grow indefinitely.

4   Minify and output *part1* .

5   Save *part2* (which is added to the start of the next chunk).

The process runs again for each incoming chunk.

That's your next coding challenge—*if you're willing to accept it!*

Worker Threads

Chapter 9 discussed how Node.js applications run on a single thread. Assume a user could trigger a complex, ten-second JavaScript calculation in your Express application. The calculation would become a bottleneck that halted processing for all users. Your application can't handle any requests or run other functions until it completes.

 Asynchronous Calculations

Complex calculations that process data from a file or database may be less problematic, because each stage runs asynchronously as it waits for data to arrive. Processing occurs on separate iterations of the event loop.

However, long-running calculations written in JavaScript alone—such as image processing or machine-learning algorithms—will hog the current iteration of the event loop.

One solution is worker threads[25]. These are similar to browser web workers[26] and launch a JavaScript process on a separate thread. The main and worker thread can exchange messages to trigger or terminate processing.

25. https://nodejs.org/dist/latest/docs/api/worker_threads.html
26. https://developer.mozilla.org/docs/Web/API/Web_Workers_API

Workers and Event Loops

Workers are useful for CPU-intensive JavaScript operations, although the main Node.js event loop should still be used for asynchronous I/O activities.

The example code has a *worker* project[27] that exports a *diceRun()* function in *lib/dice.js* . This throws any number of N-sided dice a number of times and records a count of the total score (which should result in a Normal distribution curve[28]):

```
// dice throwing
export function diceRun(runs = 1, dice = 2, sides = 6) {

  const stat = [];

  while (runs > 0) {

    let sum = 0;
    for (let d = dice; d > 0; d--) {
      sum += Math.floor( Math.random() * sides ) + 1;
    }

    stat[sum] = (stat[sum] || 0) + 1;

    runs--;
  }

  return stat;

}
```

The code in *index.js* starts a process that runs every second and outputs a message:

---

27. https://github.com/spbooks/ultimatenode1/tree/main/ch12/worker
28. https://en.wikipedia.org/wiki/Normal_distribution

```
// run process every second
const timer = setInterval(() => {
  console.log('  another process');
}, 1000);
```

Two dice are then thrown one billion times using a standard call to the
diceRun() function:

```
import { diceRun } from './lib/dice.js';

// throw 2 dice 1 billion times
const
  numberOfDice = 2,
  runs = 999_999_999;

const stat1 = diceRun(runs, numberOfDice);
```

This halts the timer, because the Node.js event loop can't continue to the next
iteration until the calculation completes.

The code then tries the same calculation in a new *Worker* . This loads a script
named *worker.js* and passes the calculation parameters in the *workerData*
property of an options object:

```
import { Worker } from 'worker_threads';

const worker = new Worker('./worker.js', { workerData: { runs, numberOfDice } });
```

Event handlers are attached to the *worker* object running the *worker.js*
script so it can receive incoming results:

```
// result returned
worker.on('message', result => {
  console.table(result);
});
```

... and handle errors:

```
// worker error
worker.on('error', e => {
  console.log(e);
});
```

... and tidy up once processing has completed:

```
// worker complete
worker.on('exit', code => {

  // tidy up

});
```

The *worker.js* script starts the *diceRun()* calculation and posts a message to the parent when it's complete—which is received by the *"message"* handler above:

```
// worker thread
import { workerData, parentPort } from 'worker_threads';
import { diceRun } from './lib/dice.js';

// start calculation
const stat = diceRun( workerData.runs, workerData.numberOfDice );

// post message to parent script
parentPort.postMessage( stat );
```

The timer isn't paused while the worker runs, because it executes on another CPU thread. In other words, the Node.js event loop continues to iterate without long delays.

Run the project code with *node index.js* .

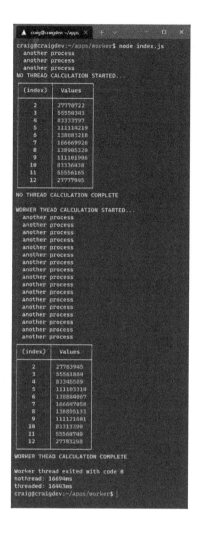

12-4. The worker output

You should note that the worker-based calculation runs slightly faster because the thread is fully dedicated to that process. Consider using workers if you encounter performance bottlenecks in your application.

View the demonstration video[29] to see the code in action.

29. https://spnt.co/nodevid26

Child Processes

It's sometimes necessary to call applications that are either not written in Node.js or have a risk of failure.

 A Real-world Example

I worked on an Express application that generated a fuzzy image hash used to identify similar graphics. It ran asynchronously and worked well—until someone uploaded a malformed GIF containing a circular reference (animation frameA referenced frameB which referenced frameA).

The hash calculation never ended. The user gave up and tried uploading again. And again. And again. The whole application eventually crashed with memory errors.

The problem was fixed by running the hashing algorithm in a child process. The Express application remained stable because it launched, monitored, and terminated the calculation when it took too long.

The child process API[30] allows you to run sub-processes that you can monitor and terminate as necessary. There are three options:

- *spawn* : spawns a child process.
- *fork* : a special type of spawn that launches a new Node.js process.
- *exec* : spawns a shell and runs a command. The result is buffered and returned to a callback function when the process ends.

Unlike worker threads, child processes are independent from the main Node.js script and can't access the same memory.

---

30. https://nodejs.org/dist/latest/docs/api/child_process.html

Clusters

Is your 64-core server CPU under-utilized when your Node.js application runs on a single core? **Clusters**[31] allow you to fork any number of identical processes to handle the load more efficiently.

The initial primary process can fork itself—perhaps once for each CPU returned by `os.cpus()`. It can also handle restarts when a process fails, and broker communication messages between forked processes.

Clusters work amazingly well, but your code can become complex. Simpler and more robust options include:

- process managers such as PM2[32], which offer an automated Cluster Mode
- a container management system such as Docker or Kubernetes

Both can start, monitor, and restart multiple isolated instances of the same Node.js application. The application will remain active even when one fails.

 Write Stateless Applications

> This was mentioned in Chapter 3, but it's worth reiterating: *make your application stateless to ensure it can scale and be more resilient*. It should be possible to start any number of instances and share the processing load.

Exercises

Browse the Readline API documentation[33] and write a small console application that prompts the user for their name before displaying a "Hello <name>" greeting.

---

31. https://nodejs.org/dist/latest/docs/api/cluster.html
32. https://pm2.keymetrics.io/
33. https://nodejs.org/dist/latest/docs/api/readline.html

Examine the Performance hooks API documentation[34] to discover how you can monitor and improve code efficiency. The worker threads code (from the "Worker Threads" section above) illustrates basic use of performance marks and >measurements.

For big bonus points, improve the stream example (from the "Streams" section) to parse incoming data chunks, as discussed above.

Summary

This chapter has provided a sample of the more useful Node.js APIs, but I encourage you to browse the documentation and discover them for yourself. The documentation is generally good and shows simple examples, but it can be terse in places. Where necessary, search for more thorough tutorials on SitePoint.

The next chapter will build on your Node.js knowledge to develop a real-time, multiuser quiz application.

Quiz

1. The *process* object provides:

    a. a way to launch a new thread
    b. information about your application and environment
    c. tools to manage application execution
    d. all of the above

2. The File System API is named:

    a. *filesystem*
    b. *file-system*
    c. *fsystem*
    d. *fs*

---

34. https://nodejs.org/dist/latest/docs/api/perf_hooks.html

3. Objects that emit events:

a. are instances of the `EventEmitter` class
b. run an `emit()` method
c. provide `on()` event handlers
d. all of the above

4. A Node.js *stream* provides:

a. data processing on smaller more manageable chunks
b. custom event management
c. processing threads management
d. asynchronous function management

5. Worker threads are best used to run:

a. asynchronous I/O activities
b. CPU-intensive JavaScript operations
c. non-Node.js applications
d. child processes

Example Real-time Multiplayer Quiz: Overview

13

This chapter demonstrates a real-time multiplayer quiz written in Node.js. The application is a step up from the simpler, self-contained examples shown in previous chapters. It has a more complex architecture, but it isn't using any modules or techniques you haven't seen before. I recommend that you progress through the explanations at your own pace and examine the code in an editor so you can follow what's happening.

The game allows any player to start a new quiz using their own configuration options—such as the number of questions, scoring, time limits, and so on. Any number of other players can join that quiz using a unique code.

Any number of quiz games can be running concurrently. Players may be connected to different HTTP and WebSocket servers, which must keep themselves synchronized as events occur.

This chapter describes how to run and play the game. The following chapters will cover these topics:

- the application's architecture (Chapter 14)
- the Express code (Chapter 15)
- the WebSocket code (Chapter 16)

Source Code

The source code is provided in the *code/ch13/nodequiz/* directory[1], although you may find it more practical to pull the repository from https://github.com/craigbuckler/nodequiz using the following Git command:

```
git clone https://github.com/craigbuckler/nodequiz
```

Quizzing Quick Start

The application uses Docker[2] and Docker Compose[3] to download and run

---

[1]. https://github.com/spbooks/ultimatenode1/tree/main/ch13/nodequiz

Node.js and database servers.

 What is Docker?

> Docker provides a way to quickly install, configure, and run
> applications such as databases. Each application launches in an
> isolated environment known as a **container**. It behaves a little like a
> Linux virtual machine, but it's lightweight and requires no ongoing
> maintenance.
>
> Docker Compose can run any number of containerized applications
> from a single command. This makes it ideal for managing web
> application dependencies, and it behaves identically on all
> platforms—whether you're using Windows, macOS, or Linux. A
> similar environment can also be deployed to a production server.
>
> The *Docker for Web Developers*[4] book and a video course[5] are
> available from SitePoint if you want to learn more.

Once you've installed Docker[6], navigate to the project root directory
(`nodequiz`) and start the application in development mode with `docker-compose up` .

All software dependencies download and initialize, so the first run can take
several minutes. The terminal shows a log of database and server activities.

Once started, access the quiz in a browser at http://quiz.localhost/.

[2.] https://docs.docker.com/get-docker/

[3.] ttps://docs.docker.com/compose/install/

[4.] https://www.sitepoint.com/premium/books/docker-for-web-developers/

[5.] https://www.sitepoint.com/premium/courses/docker-for-web-developers-3111

[6.] https://dockerwebdev.com/tutorials/install-docker/

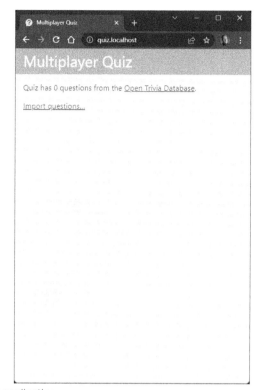

13-1. Starting the quiz application

You must import some questions before starting a quiz, so click **Import questions...** to retrieve a selection from the Open Trivia Database[7]—which is a free-to-use, user-contributed trivia question database. You're then prompted to **JOIN** an existing game.

---

7. https://opentdb.com/

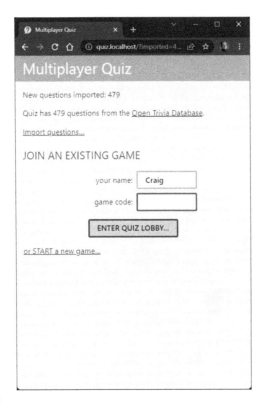

13-2. Joining the quiz

Or you can **START** a new game.

13-3. Starting a quiz

Any number of games can be running concurrently with different configurations, leading to different strategies based on time limits and whether you guess or decline to answer a question.

Once started, a game is assigned a unique code—such as **a23**, as shown below. Others can join this game by entering the code on the **JOIN** screen and entering the lobby.

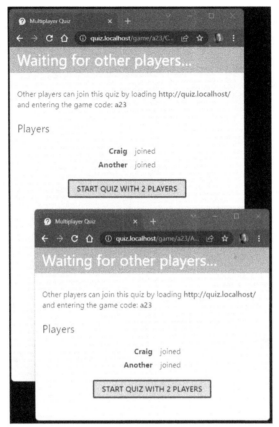

13-4. Joining the quiz

Any player can start the game, which progresses to the first question. A
countdown timer starts after the first person has answered so everyone else
must respond within the allotted time.

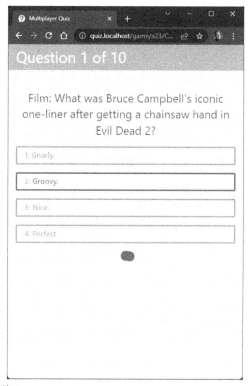

13-5. Answering a question

The score is displayed between questions and at the end of the game. A timer indicates that the next question is coming.

13-6. Viewing the quiz score

To shut down the quiz, navigate to the project root directory (*nodequiz*) in
another terminal and run *docker-compose down* .

View the demonstration video[8] to see the game running.

Summary

The quiz is developed in Node.js using a few third-party modules, vanilla ES6,
and less than 60KB of code. It's also scalable: you can add more Node.js HTTP
and WebSocket servers as traffic increases. This leads to some considerable
software engineering challenges, which we'll discuss in the next chapter.

---

[8] https://spnt.co/nodevid23

Example Real-time Multiplayer Quiz: Architecture

14

This chapter describes the quiz application's architecture and dependencies. It does get complex, so you can skip it if you'd rather concentrate on the Node.js and Express parts (Chapter 15) and WebSocket code (Chapter 16). That said, technical decisions described in those chapters are based on the architecture, so it's good to understand the basics.

Why Develop Using Multiple Servers?

You *could* develop and run the quiz on a single server running a database and a single Node.js application that launches both the HTTP and WebSocket servers. It would be easier to develop, and it would support dozens of concurrent users. However, problems will arise as traffic grows. If your application crashes, it fails for everyone, and *it's difficult to scale the quiz*:

▓ Node.js applications run on a single CPU core.

Using a multi-core CPU has negligible benefit: Node.js will use one. You could use clustering[1] (see Chapter 12), but it's a considerable coding effort, and you're still limited to the number of physical CPUs.

▓ You can't launch multiple application instances.

A process manager such as PM2[2] can launch multiple isolated instances of your application on different domains and/or HTTP ports. Two players wanting to join the same quiz would have to ensure they're connected to the same instance.

The quiz therefore uses a multi-server architecture running at least seven individual stateless applications. New Node.js application instances can be started on the same server—or even different servers—and the they'll start to handle incoming traffic. A server can fail and restart without noticeable downtime.

---

[1] https://nodejs.org/api/cluster.html
[2] https://pm2.keymetrics.io/

The only reliable way to develop this application is to use an appropriate
architecture from the start.

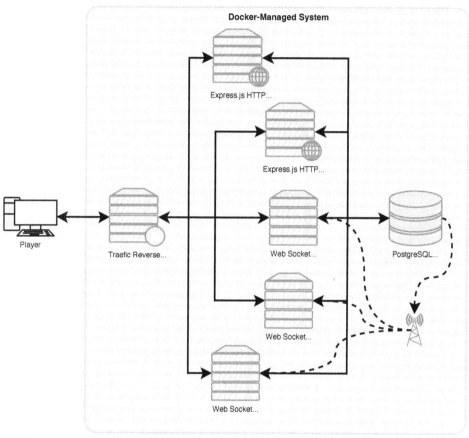

14-1. The quiz application architecture

The video for this chapter[3] and the following sections describe the setup.

1. One PostgreSQL Database Server

A single PostgreSQL database server implements a *quiz* database with the
following data tables:

- *question* : question text

---

3. https://spnt.co/nodevid24

- *answer* : answer text with correct/incorrect flags
- *game* : individual game instances and configurations
- *player* : players connected to each game
- *pubsub* : data shared to all WebSocket servers when specific events occur

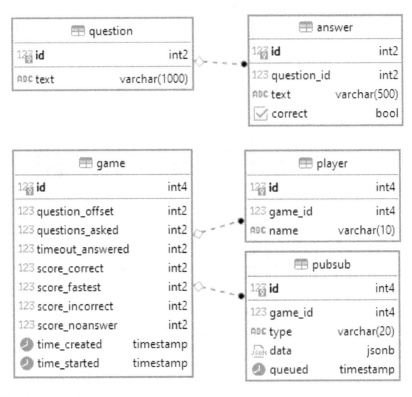

14-2. The database schema

The database guarantees data integrity using constraints defined in the schema. For example:

- it's impossible to add two questions with the same *text*
- changing the *id* of a question automatically updates the *question_id* for associated *answer* records
- deleting a *game* record automatically deletes *player* and *pubsub* records associated with that game

PostgreSQL was chosen for the project because it offers a publisher–pubscriber (or *pub–sub*) service. It's possible to trigger events when an application changes database data (publishes), which can notify all subscribers. This means we don't require a dedicated pub–sub solution as another dependency.

The `.env` file in the project root configures the database connectivity credentials. The `/db/001-quiz.sql` file creates the table schema when PostgreSQL is first launched.

Note that Express and WebSocket applications use the same Node.js module defined at `libshared/quizdb.js` for all database activities.

2. Two Express HTTP Web Servers

An Express application handles:

- importing questions from the Open Trivia Database
- allowing a player to start a new game with specific defaults
- allowing other players to join that game

It serves all the client-side HTML, CSS, and JavaScript files. Eventually, all users on the same game end up at the URL `/game/<gameCode>/<playerName>` where:

- `<gameCode>` is a unique code for a specific quiz game
- `<playerName>` is the player's name

At this point, each user connects to a WebSocket server that controls further interaction—such as starting and answering questions.

The system starts two isolated instances of the web server. This offers improved performance and redundancy: if one web server fails, the other remains active while the first restarts. An incoming HTTP request from any user can be handled by either server.

The code is defined in the `web` directory. The `.env` configuration file and `libshared` directory also provide code shared across all components.

3. Three WebSocket Servers

A WebSocket server uses the ws library to handle:

- the initial connection from a player joining a game
- starting the game for all connected players
- sending questions and answers to all players
- responding to player choices and keeping score
- disconnecting users when they leave or the game completes

The system starts three isolated instances of the WebSocket server. This offers improved performance and some redundancy. If a server fails, a game should continue for those connected to a different server.

When a player connects, they remain connected to the same WebSocket server throughout the duration of their session. However, two players on the same quiz game *could* connect to different WebSocket servers!

Games are kept in sync using the pub–sub functionality in PostgreSQL:

1. When an event occurs on one WebSocket server, such as the user answering a question, that server inserts a new record into the `pubsub` table.

2. PostgreSQL activates a trigger, which sends an event containing the new data to all WebSocket servers (the subscribers) running the same instance of a specific game.

3. Each WebSocket server (including the one that originally received the request) runs a handler that acts on the incoming event data.

The code is defined in the `ws` directory. The `.env` configuration file and

Libshared directory also provide code shared across all components.

4. One Traefic Load Balancer

Traefic[4] is a reverse proxy and load balancer that directs all incoming HTTP and WebSocket requests to the appropriate (and least-busy) server.

When the application is running, the Traefic dashboard can be accessed and monitored at http://localhost:8080/.

5. Adminer Database Client

An (optional) Adminer database client is launched in development mode so you can examine PostgreSQL table data. Access it at http://adminer.localhost/ with the following credentials:

- **System**: *PostgreSQL*
- **Server**: *dbserver* (or *host.docker.internal* or your PC's IP address)
- **Username**: *quizuser*
- **Password**: *quizpass*
- **Database**: *quiz*

If you'd rather use your preferred client application to access the database, enter *localhost* as the **Server** name. Popular options including Beekeeper Studio[5], DBeaver[6], and Postbird[7] should be compatible.

Docker Development Environment

Installing, configuring, and launching all seven applications (eight with Adminer) on a single server wouldn't leave much time for development! Fortunately, the whole environment can be managed with Docker so it starts

---

[4.] https://traefik.io/traefik/
[5.] https://beekeeperstudio.io/
[6.] https://dbeaver.io/
[7.] https://github.com/paxa/postbird

in a few seconds and still supports live Node.js restarts using `nodemon` .

For this reason, Docker is the only software dependency you need to install. *Even the Node.js runtime is managed by Docker.*

A full Docker tutorial is beyond the scope of this book, but the `web` and `ws` directories have `Dockerfile` configurations (`web.Dockerfile` and `ws.Dockerfile`), which tell Docker how to build and run the Express and WebSocket applications from a lightweight Node.js 16 Alpine Linux base.

Both create a Docker **image**. You can think of it like a disk image containing all the files, libraries, and executables required to run an application.

You can start any number of Docker image instances. A running instance is known as a **container**. Think of it as an isolated Linux Virtual Machine that's running a single executable such as a database or Node.js application.

Launching a container requires a single `docker run` command. Fortunately, Docker Compose can manage and run all containers using a development environment configuration defined in `docker-compose.yml` . This does the following:

- Declares all containers, replicas, and restart policies.

- Defines all environment variables from the `.env` file.

- Attaches disk storage volumes so there's no need to re-initialize the database on every launch. It also mounts the `libshared` modules directory in both the `web` and `ws` projects.

- Overrides some `Dockerfile` settings to use `nodemon` and launch Node.js debugging servers.

- Connects all containers to the same internal Docker network.

Configures the Traefic load balancer.

Start the whole environment in development mode from the project's root directory:

```
docker-compose up
```

The terminal shows a live activity log and any errors. `nodemon` restarts the `web` and `ws` applications whenever a JavaScript file is changed.

To gracefully shut down all applications, run the following command in another terminal from the project root:

```
docker-compose down
```

Docker Production Environment

`docker-compose-production.yml` defines production-level settings, so the quiz application can be run on a live server. The configuration is simpler, because there's no need to override `Dockerfile` settings or launch Adminer.

Start the application in production mode with:

```
docker-compose -f ./docker-compose-production.yml up
```

 Is Docker Compose Suited to Production?

Probably not. It's not efficient to run PostgreSQL in a container, and there are better options such as Docker Swarm[8] and Kubernetes[9] to manage containers across multiple servers. But that's beyond the scope of this and most Docker books!

---

8. https://docs.docker.com/engine/swarm/
9. https://kubernetes.io/

Summary

Setting up a development environment is complex, but the choices you make at the start can affect the long-term success of your project. We're now in a good position to create a (mostly) stateless application, starting with the Express server in the next chapter.

Chapter

Example Real-time Multiplayer Quiz: Express Code

15

The Express part of the quiz application:

- imports questions from the Open Trivia Database
- allows a player to create and start a new game with specific defaults
- allows other players to join that game

It serves all the client-side HTML, CSS, and JavaScript files. Eventually, users on the same game end up at the URL *`/game/<gameCode>/<playerName>`* , where the WebSocket server (see Chapter 16) takes over and controls the gameplay.

Docker starts two isolated HTTP servers and a single request could be directed to either by the Traefic load balancer. Even two requests from the same user on the same page—such as a CSS and JavaScript file—could be delivered by different servers. This is rarely an issue, because the web is stateless by default: the application avoids storing state on one server that wouldn't be available on the other.

Before we delve into the Express code, we'll take a look at the database code.

Database Library

PostgreSQL database connectivity is handled by the Node.js *`pg`* library[1] (see the documentation[2]). This is loaded in the *`libshared/quizdb.js`* module, which provides a selection of functions to *`INSERT`* , *`UPDATE`* , *`SELECT`* , and *`DELETE`* records in the *`quiz`* database. The same module is used by both the *`web`* and *`ws`* servers.

The code initially imports the *`pg`* library and defines integer type parsers. By default, *`pg`* returns all record fields as strings, so a parser can convert it to the correct type:

```
import pg from 'pg';
```

---

[1] https://www.npmjs.com/package/pg
[2] https://node-postgres.com/

```
// data type parsers
pg.types.setTypeParser(pg.types.builtins.INT2, v => parseInt(v, 10));
pg.types.setTypeParser(pg.types.builtins.INT4, v => parseInt(v, 10));
pg.types.setTypeParser(pg.types.builtins.INT8, v => parseFloat(v));
```

The code then defines a "connection pool"[3] using the environment variable defaults:

```
const pool = new pg.Pool({
  host: process.env.POSTGRES_SERVER,
  port: process.env.POSTGRES_PORT,
  database: process.env.POSTGRES_DB,
  user: process.env.POSTGRES_QUIZUSER,
  password: process.env.POSTGRES_QUIZPASS
});
```

A **pool** provides a reusable set of database connection clients you can check out, use, release, and reuse. This has benefits including:

- There's no initial handshake delay when a client is reused.
- Each client is a separate connection to the database. Unlike a single connection, they can make simultaneous requests.

Here's a basic parameterized SQL query example that returns all records from the *question* table with an *id* between 1 and 10 using one of the pool connections:

```
// DB connection
const client = await pool.connect();

try {
  // fetch all questions with ids between 1 and 10
  const result = await client.query(
    'SELECT * FROM question WHERE id >= $1 AND id <= $2;',
    [1, 10]
  );
```

[3.] https://node-postgres.com/features/pooling

```
}
catch(err) {
  console.log(err);
}
finally {
  // release client
  client.release();
}
```

The SQL *SELECT* string references *$1* and *$2* , which are substituted with values in the first and second elements in the array. An array of row objects is returned when the query executes successfully.

Creating individual SQL commands can be cumbersome, and it's easy to miss or transpose array parameters. The *libshared/quizdb.js* module has private *dbSelect()* , *dbInsert()* , *dbUpdate()* , and *dbDelete()* functions, which make development easier. For example, the public *playerCreate()* function is used when adding a new *player* record for a specific game:

```
// create a new player
export async function playerCreate( game_id, name ) {

  return await dbInsert({
    table: 'player',
    values: { game_id, name },
    return: 'id'
  });

}
```

This calls the private *dbInsert()* function with a *table* name, a *values* object containing name/value pairs, and a *return* to fetch the *id* of the added record. The *dbInsert()* function returns the added *id* or *false* when an error occurs:

```
// database INSERT
// pass object: { table: <tablename>, values: <{ n1: v1,... }>, return: <field> }
async function dbInsert(ins) {
```

```
const
  ret = ins.return ? ` RETURNING ${ ins.return }` : '',
  key = Object.keys( ins.values ),
  sym = key.map( (v,i) => `$$${i + 1}` ),
  sql = `INSERT INTO ${ ins.table } (${ key.join() }) VALUES(${ sym.join() })
  ↪${ ret };`,
  client = ins.client || await pool.connect();

let success = false;

try {

  // run insert
  const i = await client.query(sql, Object.values( ins.values ));

  // successful?
  success = i.rowCount === 1;

  // return value?
  if (success && ins.return) {
    success = i.rows[0][ ins.return ];
  }

}
catch(err) {
}
finally {
  if (!ins.client) client.release();
}

return success;

}
```

The *const* values at the top are responsible for creating the SQL string:

```
INSERT INTO player (game_id, name) VALUES ($1, $2) RETURNING id;
```

There are four things to note here:

▨ *key* defines an array of property names extracted from the *values* object.

▦   *sym* defines an array of *$1* to *$N* strings, which match the number of items in the *key* array.

▦   The property values from *values* are passed to the SQL query using *Object.values(ins.values)* .

▦   The calling function can pass its own *pool.connect()* object. This is necessary when it's running a series of updates in a database transaction.

The private *dbUpdate()* method is similar, although it also receives a *where* object with name/value pairs to create an SQL string, such as:

```
UPDATE game SET time_started=$1 WHERE game_id=$2;
```

The function ensures the names and values resolve correctly:

```
// database UPDATE
// pass object: { table: <tablename>, values: <{ n1: v1,... }>,
// where: <{ n1: v1,... }> }
async function dbUpdate(upd) {

  const
    sym = [...Object.values( upd.values ), ...Object.values( upd.where )],
    vkey = Object.keys( upd.values ),
    val = vkey.map( (k, i) => `${ k }=$$${ i + 1 }` ),
    ckey = Object.keys( upd.where ),
    cond = ckey.map( (k, i) => `${ k }=$$${ i + val.length + 1 }` ),
    sql = `UPDATE ${ upd.table } SET ${ val.join() } WHERE ${ cond.join() };`,
    client = upd.client || await pool.connect();

  let updated = 0;

  try {

    // run update
    const u = await client.query(sql, sym);

    // successful?
    updated = u.rowCount;
```

```
    }
    catch(err) {
    }
    finally {
      if (!upd.client) client.release();
    }

    return updated;

  }
```

Record deletion SQL is simpler. For example:

```
 DELETE FROM game WHERE id=$1;
```

Therefore, so is the *dbDelete()* function:

```
 // database delete
 // pass object: { table: <tablename>, where: <{ n1: v1,... }> }
 // logical AND is used for all where name/value pairs
 async function dbDelete(del) {

   const
     key = Object.keys( del.values ).map((v, i) => `${ v }=$${ i+1 }`),
     sql = `DELETE FROM ${ del.table } WHERE ${ key.join(' AND ') };`,
     client = del.client || await pool.connect();

   let deleted = false;

   try {

     // run delete
     const d = await client.query(sql, Object.values( del.values ));
     deleted = d.rowCount;

   }
   catch(err) {
   }
   finally {
     if (!del.client) client.release();
   }
```

```
    return deleted;

}
```

Finally, *dbSelect()* is the simplest function of all, since you must specify your own *sql* string and array of arguments:

```
// database SELECT
// pass SQL string and array of parameters
async function dbSelect(sql, arg = []) {

  const client = await pool.connect();

  try {
    const result = await client.query(sql, arg);
    return result && result.rows;
  }
  catch(err) {
    console.log(err);
  }
  finally {
    client.release();
  }

}
```

The reason is that SQL *SELECT* queries can be varied and complex. Some database libraries provide object–relational mapping (ORM) methods to build SQL query strings, but this would have been overkill for this project!

Question Database Initialization

Data is downloaded from the Open Trivia Database—a free-to-use repository of user-contributed questions and answers with a REST API[4]. Questions and their associated answers are stored in the *question* and *answer* database tables. This action can be initiated by a user when the quiz home page is accessed for the first time.

[4.] https://opentdb.com/api_config.php

Initializing Data on Application Start?

The application could initialize the questions when the web server starts. However, any number of application instances can be launched and each would attempt to load questions. Making it into a user request ensures only one server will load questions at a time.

The `.env` file defines environment variables—including database credentials and `QUIZ_QUESTIONS_MAX=500` —to limit the number of imported questions. The `web/index.js` entry script loads modules and configures the Express server:

```
// Express
import express from 'express';
import compression from 'compression';

// modules
import { questionCount, gameCreate, gameFetch } from './libshared/quizdb.js';
import { questionsImport } from './lib/questionsimport.js';
import * as libId from './libshared/libid.js';

// configuration
const cfg = {
  dev: ((process.env.NODE_ENV).trim().toLowerCase() !== 'production'),
  port: process.env.NODE_PORT || 8000,
  domain: process.env.QUIZ_WEB_DOMAIN,
  wsDomain: process.env.QUIZ_WS_DOMAIN,
  title: process.env.QUIZ_TITLE,
  questionsMax: parseInt(process.env.QUIZ_QUESTIONS_MAX, 10)
};

// Express initiation
const app = express();

// use EJS templates
app.set('view engine', 'ejs');
app.set('views', 'views');

// GZIP
app.use(compression());
```

```
// body parsing
app.use(express.urlencoded({ extended: true }));
```

By default, the home page `/` route fetches the number of questions in the database using the `questionCount()` function in `libshared/quizdb.js` (see the `else` block):

```
// home page
app.get('/', async (req, res) => {

  if (typeof req.query.import !== 'undefined') {

    // import new questions and redirect back
    res.redirect(`/?imported=${ await questionsImport() }`);

  }
  else {

    // home page template
    res.render('home', {
      title: cfg.title,
      questions: await questionCount(),
      questionsMax: cfg.questionsMax,
      imported: req.query?.imported || null
    });

  }

});
```

This count and `questionsMax` is passed to an HTML view at `web/views/home.ejs`. It shows a link to the home page with an `/?import` query string when further questions can be loaded:

```
<% if (questions < questionsMax) { %>

  <p><a href="/?import">Import questions&hellip;</a></p>

<% } %>
```

When clicked, it reloads the home page with an `?import` query string, which triggers the `if` block above. This executes `questionsImport()` in `web/lib/questionsimport.js` and returns the number of questions imported.

The code then redirects back to the home page with an `?imported=N` query string, which shows the number of imported questions. Assuming there's at least one question in the database, the `web/views/home.ejs` view displays the `START` and `JOIN` game options:

```
<% if (questions) { %>

  <section class="tabs">
    <article id="new">
      <h2>START A NEW GAME</h2>
      <!-- more code -->

<% } %>
```

 Why Does the Number of Imported Questions Vary?

The Open Trivia Database API returns a random set of questions. Some may be duplicates of previously imported questions, but the database's `question.text` field has a `UNIQUE` flag to ensure a question can only be added once.

The `questionsImport()` function is a little long, so examine `web/lib/questionsimport.js` in an editor. It uses a series of promise-based functions to make up to ten concurrent calls to the Open Trivia API at https://opentdb.com/api.php with `Promise.allSettled()`. Data is fetched using the `node-fetch` module.

Native Node.js Fetch()

Deno usefully implements the browser Fetch API, so you can use it
in a server application. A similar Fetch API arrived in Node.js version
18, but it's experimental. A third-party module is used here for
backward compatibility.

The Open Trivia API returns JSON data such as:

```
{
  "response_code": 0,
  "results": [
    {
      "category": "History",
      "type": "multiple",
      "difficulty": "medium",
      "question": "The crown of the Empire State Building was originally built
      ↪for what purpose?",
      "correct_answer": "Airship Dock",
      "incorrect_answers": [
        "Lightning Rod",
        "Antennae",
        "Flag Pole"
      ]
    },
    {
      "category": "Entertainment: Cartoon & Animations",
      "type": "multiple",
      "difficulty": "easy",
      "question": "Which of these is NOT a Disney cartoon character?",
      "correct_answer": "Daffy Duck",
      "incorrect_answers": [
        "Donald Duck",
        "Daisy Duck",
        "Scrooge McDuck"
      ]
    },
    {
      "category": "History",
      "type": "multiple",
```

```
      "difficulty": "hard",
      "question": "What was the original name of New York City?",
      "correct_answer": "New Amsterdam",
      "incorrect_answers": [
        "New London",
        "New Paris",
        "New Rome"
      ]
    }
  ]
}
```

This is converted to JavaScript values, formatted, and each question/answer set is added to the database using a call to the _questionAdd(question, answer)_ function in _libshared/quizdb.js_ . Each question and answer set is inserted within a database transaction so that, if any SQL INSERT operation fails, they all fail:

```
// add a new question and answer set
export async function questionAdd(question, answer) {

  const client = await pool.connect();
  let commit = false;

  try {

    // new transaction
    await client.query('BEGIN');

    // add question
    const qId = await dbInsert({
      client,
      table: 'question',
      values: {
        text: question
      },
      return: 'id'
    })

    if (qId) {
```

```
    // insert answers in sequence
    let inserted = 0;
    for (let item of answer) {

      const a = await dbInsert({
        client,
        table: 'answer',
        values: {
          question_id: qId,
          text: item.text,
          correct: item.correct
        }
      });

      if (a) inserted++;

    }

    // answers added?
    commit = inserted === answer.length;

  }

}
catch(err) {
}
finally {

  // commit or rollback transaction
  if (commit) {
    await client.query('COMMIT');
  }
  else {
    await client.query('ROLLBACK');
  }

  client.release();
}

return commit;

}
```

Sequential Database INSERTs

The code could run multiple database `INSERT` commands in a short period. This is faster, but question and answer IDs would appear in a seemingly random order in the database tables. For example, the `question` record with an `id` of 1 could have associated `answer` records with the `id`s 17, 22, 52, and 54.

This isn't a problem for an indexed database, but it can make the tables more difficult to read during development! For this reason, questions and answers are inserted sequentially, one at a time. It also means that ordering by `answer.id` returns an alphabetically ordered list without requiring an additional `answer.order` field.

Starting a New Game

The `web/views/home.ejs` template defines an HTML form to configure and start new games:

```
<form action="/newgame/" method="post">

  <div class="formgrid">

    <label for="namenew">your name:</label>
    <div><input type="text" name="name" id="namenew" value="" minlength="1"
    maxlength="10" pattern="[A-Za-z0-9]{1,10}" required /></div>

    <label for="questions_asked">number of questions:</label>
    <div><input type="number" name="questions_asked" id="questions_asked"
    value="10" min="1" max="50" required /></div>

    <label for="timeout_answered">time limit after first answer:</label>
    <div><input type="number" name="timeout_answered" id="timeout_answered"
    value="5" min="5" max="60" required /> seconds</div>

    <label for="score_correct">score for correct answer:</label>
    <div><input type="number" name="score_correct" id="score_correct" value="1"
    min="-100" max="100" required /> points</div>
```

```
<label for="score_fastest">bonus for fastest player:</label>
<div><input type="number" name="score_fastest" id="score_fastest" value="1"
min="-100" max="100" required /> points</div>

<label for="score_incorrect">score for incorrect answer:</label>
<div><input type="number" name="score_incorrect" id="score_incorrect"
value="-1" min="-100" max="100" required /> points</div>

<label for="score_noanswer">score for no answer:</label>
<div><input type="number" name="score_noanswer" id="score_noanswer" value="0"
min="-100" max="100" required /> points</div>
</div>

<button>ENTER QUIZ LOBBY&hellip;</button>

</form>
```

The form HTTP POSTs data to the /newgame/ URL, which is handled by the
route defined in web/index.js :

```
// create a new game
app.post('/newgame', async (req, res) => {

  const
    gameId = await(gameCreate( req.body )),
    playerName = libId.clean( req.body.name );

  if (gameId === null) {

    // game creation error?
    res.status(500).render('error', {
      title: cfg.title,
      error: 'Game could not be started?'
    });

  }
  else {

    // redirect to game page using slug and user name
    res.redirect(`/game/${ libId.encode( gameId ) }/${ playerName }`);
```

```
  }

});
```

The code calls the `gameCreate()` function in `libshared/quizdb.js` and passes the `req.body` object containing the form data. This inserts a new record into the database `game` table and returns its `id` —by calling the private `dbInsert()` function (shown above in the "Database Library" section):

```
// create a new game
export async function gameCreate(data) {

  const qCount = await questionCount();

  return await dbInsert({
    table: 'game',
    values: {
      question_offset : Math.floor( Math.random() * qCount ), // random start q
      questions_asked : clamp(1, data.questions_asked, 50),
      timeout_answered: clamp(5, data.timeout_answered, 60),
      score_correct   : clamp(-100, data.score_correct, 100),
      score_fastest   : clamp(-100, data.score_fastest, 100),
      score_incorrect : clamp(-100, data.score_incorrect, 100),
      score_noanswer  : clamp(-100, data.score_noanswer, 100)
    },
    return: 'id'
  });

}
```

Note the following:

▪ Each *game* record has a unique `id` integer which identifies the game.

  The number can become long and is easy to guess. If you're currently playing game `99`, you could try joining game `100` or `101` and have a high success rate.

  For this reason, game IDs are encrypted into a string using *encode()* and decode() in *libshared/libid.js* . This string also avoids using similar-looking characters such as zero and uppercase "o" or one and uppercase "i".

A player can then tell others to join game *a23* rather than game *1* .

- *clamp()* is a private function that ensures a value is between a lower and upper limit:

```
// return integer between low and high values
function clamp(min = 0, value = 0, max = 0) {

    return Math.max(min, Math.min(parseInt(value || '0', 10) || 0,
    ↪max));

}
```

- *game.question_offset* defines the starting question. It's set to a random number between 0 and the number of database questions.

- *game.time_created* is automatically set to the date/time the game was created by the database (*time_created timestamp NOT NULL DEFAULT NOW()*).

- *game.time_started* is initially *NULL* , but is eventually set to the date/time the game is started. This value is checked when you join a game to ensure players can't jump in mid-way through a quiz.

Assuming a game record is created, the browser redirects the user to the URL */game/<gameCode>/<playerName>* —such as */game/a23/Craig* . A failure shows a message using the view at *web/views/error.ejs* .

Joining a Game

The *web/views/home.ejs* template also defines an HTML form for joining a game that HTTP POSTs the user's name and game code to the */joingame/*

route:

```
<form action="/joingame/" method="post">

  <div class="formgrid">

    <label for="namejoin">your name:</label>
    <div><input type="text" name="name" id="namejoin" value="" minlength="1"
    maxlength="10" pattern="[A-Za-z0-9]{1,10}" required /></div>

    <label for="slug">game code:</label>
    <div><input type="text" name="slug" id="slug" value="" minlength="3"
    maxlength="8" autocomplete="off" required /></div>

  </div>

  <button>ENTER QUIZ LOBBY&hellip;</button>

</form>
```

The form HTTP POSTs data to the `/joingame/` URL, which is handled by the route defined in `web/index.js`:

```
// join an existing game
app.post('/joingame', (req, res) => {

  // redirect to game page using slug and user name
  res.redirect(`/game/${ libId.clean( req.body.slug ).toLowerCase() || 'x' }/${
  ↳ libId.clean( req.body.name ) }`);

});
```

This receives the data, cleans the strings, and redirects the user to the URL `/game/<gameCode>/<playerName>` —such as `/game/a23/Craig`.

Quiz Page

All players starting or joining a game reach the URL `/game/<gameCode>/<playerName>`, where:

　　<gameCode> is the unique code for a specific quiz game

　　<playerName> is a player's name

This is handled by the Express routing function at web/index.js:

```
// game page
app.get('/game/:slug/:name', async (req, res) => {

  // get game ID and player name
  const
    slug = req.params.slug,
    gameId = libId.decode( slug ),
    game = gameId === null ? null : await gameFetch( gameId ),
    gameValid = game && gameId === game.id,
    playerName = libId.clean( req.params.name ) || 'Player';

  if (gameValid && game.time_started === null) {

    // game open for players
    res.render('game', {
      domain: cfg.domain,
      wsDomain: cfg.wsDomain,
      slug,
      title: cfg.title,
      game,
      playerName
    });

  }
  else {

    // game has been started or is invalid
    const url = `${ cfg.domain }/game/${ slug }`;

    res.status(gameValid ? 403 : 404).render('error', {
      title: cfg.title,
      error: gameValid ? `You were too late to join the game at ${ url }` : `The
      ↪game at ${ url } is not valid. Did you enter it correctly?`
    });

  }
```

```
  });
```

The function decodes the game code to an integer and fetches the game information from the database by calling *gameFetch()* in *Libshared/ quizdb.js* :

```
// fetch game data
export async function gameFetch( gameId ) {

  const game = await dbSelect('SELECT * FROM game WHERE id=$1;', [ gameId ]);
  return game?.[0];

}
```

Assuming the game ID is valid and the game's *time_started* value is *NULL* , the code renders the template at *web/views/game.ejs* . Configuration variables are passed to a client-side script in the template:

```
<script type="module">
window.cfg = {
  wsDomain: '<%= wsDomain %>',
  gameId: <%= game.id %>,
  playerName: '<%= playerName %>'
};
</script>
<script type="module" src="/js/main.js"></script>
```

This configures values used in the client-side script at *web/static/js/ main.js* .

When necessary, errors are shown using the template at *web/views/ error.ejs* :

- An invalid game ID returns an HTTP 404 Not found error.
- A started game (where *time_started* is not *NULL*) returns an HTTP 403 Forbidden error.

Summary

The Express part of the application illustrates how URL routes can be resolved to trigger server-side functionality.

At this point, all players joining a game have loaded the `web/views/game.ejs` template. All further quiz game processing is now handled using client-side JavaScript and WebSocket server messaging (see Chapter 16). Express has completed its job!

Example Real-time Multiplayer Quiz: WebSocket Code

Chapter

16

Chapter 11 introduced WebSockets, which establish a two-way interactive communication channel between a client browser and server.

Our quiz application starts three WebSocket servers, and there's no limit to the number of servers that could be started. However:

- A user will connect to a single server and remain connected to it throughout their session.
- Two users on the same game *could* be connected to different WebSocket servers.

Messages sent to and from the WebSocket server are typically simple strings, but we have the added challenge of coordinating messages across all servers!

Initiating a WebSocket Connection

The client-side JavaScript at *web/static/js/main.js* initiates a connection to the WebSocket server's address and sends a *gameInit* message when it's established. Note that *window.cfg.wsDomain*, *window.cfg.gameId*, and *window.cfg.playerName* are values passed by Express to the *web/views/game.ejs* template:

```
// client-side code
// handle WebSocket communication
const ws = new WebSocket( window.cfg.wsDomain );

// connect to server and send game ID and initial player name
ws.addEventListener('open', () => {
  sendMessage( 'gameInit', { gameId: window.cfg.gameId, playerName: window.cfg.
  ↪playerName } );
});

// send message
function sendMessage(type, data = null) {
  ws.send( `${ type }:${ JSON.stringify( data ) }` );
}
```

An event handler function can now receive incoming messages from the

WebSocket server:

```
// receive message
ws.addEventListener('message', e => {

  // process...

});
```

The server-side script at *ws/index.js* initializes a ws library *WebSocketServer* object and listens for new client connections and incoming messages:

```
// server
ws = new WebSocketServer({ port: cfg.wsPort, perMessageDeflate: false });

// client connected
ws.on('connection', (socket, req) => {

  console.log(`connection from ${ req.socket.remoteAddress }`);

  // message received from client
  socket.on('message', async (msg) => {

    // process...

  });

}
```

WebSocket Message Format

The quiz application uses the same format for all WebSocket messages sent by the client or server. An identifying *type* string is followed by a colon character and payload *data* in JSON format:

```
messageType:{ jsondata }
```

For example, the *gameInit* message shown above passes the game ID and player name to the server shortly after initiating the WebSocket connection:

```
gameInit:{ "gameId": "a23", "playerName": "Craig" }
```

When receiving a message from a player, the WebSocket server may perform some actions immediately. However, most messages are forwarded to *all* WebSocket servers where users are connected to the same game. Each WebSocket server (including the one that originally received the message) then process the message and, in most cases, transmits it back to its connected clients where DOM and game state updates occur.

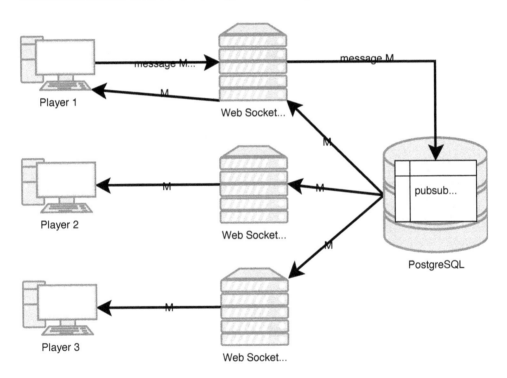

16-1. Quiz WebSocket messaging

PostgreSQL Pub–sub

Messages are broadcast to all WebSocket servers using PostgreSQL's pub–sub functionality. The *pubsub* table has the following fields:

- *id* : an auto-incrementing integer
- *game_id* : the (non-encoded) game ID integer (a foreign key for the *game.id*)
- *type* : the type of message
- *data* : the message payload in fast JSONB binary format
- *queued* : the current timestamp

When a single WebSocket server wants to share an event, it inserts a new record into the *pubsub* table using a *broadcast()* function in *Libshared/ quizdb.js* :

```js
// broadcast an event
export async function broadcast( game_id, type, data ) {

  return await dbInsert({
    table: 'pubsub',
    values: { game_id, type, data },
    return: 'id'
  });

}
```

A database trigger named *pubsub_insert_trigger* calls a *sub_insert_notify()* function whenever a record is inserted into the *pubsub* table. This executes a PostgreSQL *pg_notify()* command, which sends the record to all subscribers:

```sql
CREATE OR REPLACE FUNCTION pubsub_insert_notify()
  RETURNS trigger AS
$BODY$
  BEGIN
    PERFORM pg_notify('pubsub_insert', row_to_json(NEW)::text);
    RETURN NULL;
END;
$BODY$
  LANGUAGE plpgsql VOLATILE
  COST 100;
```

```
CREATE TRIGGER "pubsub_insert_trigger"
  AFTER INSERT ON public.pubsub FOR EACH ROW
  EXECUTE FUNCTION pubsub_insert_notify();
```

A *PubSub* object, which extends the Node.js *EventEmitter* class, is defined in *Libshared/quizdb.js* :

```
// pubsub event emitter
class PubSub extends EventEmitter {

  constructor(delay) {
    super();
  }

  async listen() {

    if (this.listening) return;
    this.listening = true;

    const client = await pool.connect();

    client.on('notification', event => {

      try {
        const payload = JSON.parse( event.payload );
        if ( payload ) {

          this.emit(
            `event:${ payload.game_id }`,
            {
              gameId: payload.game_id,
              type: payload.type,
              data: payload.data
            }
          );

        }
      }
      catch (e) {
      }

    });
```

```
    client.query('LISTEN pubsub_insert;');

  }

}
```

A single object instance named *pubsub* is instantiated and exported. The *listen()* method is called, which connects to the database and defines a handler function when a *notification* event occurs. This emits a Node.js event named *event:<gameId>* with a payload object containing the *gameId*, *type* string, and JSON-parsed *data* object:

```
export const pubsub = new PubSub();
await pubsub.listen();
```

Game instances on each server then subscribe to *event:<gameId>* event using the *pubsub.on* method:

```
import * as db from '../libshared/quizdb.js';

// abbreviated code
class Game {

  #handlerFunction = async e => await this.#eventHandler(e);

  // initialize game
  async create( gameId ) {

    this.gameId = gameId;

    // monitor incoming events
    db.pubsub.on(`event:${ this.gameId }`, this.#handlerFunction);

  }

}
```

The private *#eventHandler()* function is called when an event occurs. It receives the incoming data and can react accordingly:

```
// incoming event sent to all game servers
async #eventHandler({ gameId, type, data }) {

  console.log('Shared server event', type, data);
  // ...

  // handle server event (on all servers)
  switch (type) {
    //...
  }

}
```

Game Logic

This section provides an overview of the game logic as it progresses from joining, to starting, playing, and finishing a quiz. Note the following:

- The client-side JavaScript at *web/static/js/main.js* sends messages from a client to the WebSocket server when an event occurs—such as answering a question.

- The WebSocket server-side JavaScript at *ws/index.js* receives a message from a client and executes appropriate functionality. In most cases, messages are broadcast to all WebSocket servers running the same game. They receive the data and transmit it back to all connected clients on that game.

- The client-side JavaScript at *web/static/js/main.js* receives incoming messages and updates the DOM or game state accordingly.

  The HTML *<body>* class is set to the incoming message *type*. This can trigger CSS to show or hide specific elements according to the game state.

In some cases, an action occurring on a client does nothing until it has been received *back* from the server after it has been transmitted to all WebSocket servers and their connected clients!

Joining a Game

When a player accesses a valid game at the URL
 `/game/<gameCode>/<playerName>` —such as `/game/a23/Craig` —the client
establishes a WebSocket connection with a single server and sends a
 `gameInit` message. For example:

```
gameInit:{ "gameId": "a23", "playerName": "Craig" }
```

This triggers the `message` event handler function on a WebSocket Server,
which initiates the game (the `if` block):

```
// message received from client
socket.on('message', async (msg) => {

  // parse message
  msg = parseMessage(msg);

  // initialize player and game
  if (!player && msg.type === 'gameInit' && msg.data) {

    player = new Player();
    const pId = await player.create( msg.data.gameId, msg.data.playerName,
    ↪socket );
    if (!pId) player = null;

  }
  else {

    // pass message to game object
    msg.data = msg.data || {};
    msg.data.playerId = player.id;
    await player.game.clientMessage( msg );

  }

});
```

A new `Player` object is created using the class defined in `ws/lib/player.js` .
Its `create( gameId, playerName, socket )` method is run:

```
// initialize new player
async create( gameId, playerName, socket ) {

  // player properties
  this.name = playerName;
  this.#socket = socket;

  // initialize game
  this.gameId = gameId;
  this.game = await GameFactory( gameId );
  if ( !this.game ) return null;

  // send existing players to new player
  this.send('player', this.game.playerAll())

  // create this player
  this.id = await db.playerCreate( this.gameId, playerName );
  if ( !this.id ) return null;

  // add player to game
  this.game.playerAdd( this );

  return this.id;

}
```

 ### Why Run a create() Method?

> The `Player` class has a *constructor* function that runs when an
> instance of an object is created. Unfortunately, *constructor*
> functions can't be asynchronous, so it's necessary to run another
> method to handle initialization.

`Player` objects keep track of the user's ID, game ID, name, score, and the
WebSocket connection is used by the `send()` method to send a message to
an individual player:

```
// send message to player
```

```
  send( type = 'ws', data = {} ) {

    if (this.#socket) {
      this.#socket.send( `${ type }:${ JSON.stringify(data) }` );
    }

  }
```

Player create() passes the *gameId* to a *GameFactory()* function defined in
ws/lib/game.js :

```
// active games
const gameActive = new Map();

// create and manage active game objects
export async function GameFactory( gameId ) {

  // game instance not exists?
  if ( !gameActive.has( gameId ) ) {

    // create new game instance
    const game = new Game();
    if ( await game.create( gameId ) ) {
      gameActive.set( gameId, game );
    }

    console.log(`Game ${ gameId } added - active games on this server: ${
    ↪gameActive.size }`);

  }

  return gameActive.get( gameId ) || null;

}
```

Game objects keep track of the game state and connected players. They're
responsible for receiving a message from a single client, broadcasting that
message to all WebSocket servers, and receiving the message back again,
and sending it to all connected clients on the same game.

The *GameFactory()* function creates and returns a *new Game* object when the first player joins a specific game on each WebSocket server. This object is referenced in a *gameActive* JavaScript *Map* using the game ID integer as the reference. All subsequent players to join the same game on the same WebSocket server receive the same *Game* object.

Next, the joining client is sent a *player* message with an array of all existing player IDs and names (from the *Game* object's *playerAll()* method). When received, an *init()* function in the client-side *web/static/js/player.js* script adds each player to an HTML *<table>* and stores DOM references in a *Map* named *player* :

```
// CLIENT-SIDE CODE
const
  pList = document.getElementById('player'),
  pNum = document.getElementById('pnum'),
  player = new Map();

// add new players
export function init(pAll, showScore = false) {
  clear(pList);
  player.clear();
  pAll.forEach(p => add(p, showScore));
}

// add a new player
export function add( p, showScore = false ) {

  if (!p.id || player.has(p.id)) return;

  const item = document.createElement('tr');
  (item.appendChild(document.createElement('th'))).textContent = p.name;
  const info = item.appendChild(document.createElement('td'));
  info.textContent = showScore ? p.score || 0 : 'joined';

  const pObj = {
    name: p.name,
    node: pList.appendChild(item)
  };
  pObj.info = pObj.node.getElementsByTagName('td')[0];
```

```
    player.set(p.id, pObj);
    pNum.textContent = player.size;

}
```

The player is now added to the *player* table in the database by calling the *playerCreate()* function in *libshared/quizdb.js* :

```
// create a new player
export async function playerCreate( game_id, name ) {

  return await dbInsert({
    table: 'player',
    values: { game_id, name },
    return: 'id'
  });

}
```

Assuming the player can be inserted into the database, the *Player* object (*this*) is passed to the *Game* object's *playerAdd()* method:

```
// add player to game
async playerAdd( player, broadcast = true ) {

  // add player to this server
  this.player.set( player.id, player );

  // broadcast event
  if (broadcast) {

    await db.broadcast(
      this.gameId,
      'playerAdd',
      { id: player.id, game_id: this.gameId, name: player.name }
    );

  }

}
```

This broadcasts a *playerAdd* message with the new player's ID, game ID, and name to all WebSocket servers. These send it to all connected clients on the same game (including the joining player). When received, the *add()* function in the client-side *web/static/js/player.js* script (shown above) adds the new player to the same HTML *<table>* .

Starting a Game

After joining, any player can hit the **START QUIZ** button. This sends a *start* message to one WebSocket server, which broadcasts it to all servers and back to all clients. Each client calls a *start()* function in the client-side *web/static/js/player.js* script that shows which player started the game and initializes a five-second countdown timer using the *startTimer()* function in *web/static/js/timer.js* :

```
// CLIENT-SIDE JAVASCRIPT
// started
export function start(pId) {

  if (!player.has(pId)) return;
  player.get(pId).info.textContent = 'started game';
  startTimer();

}
```

The *Game* object on the WebSocket server (*ws/lib/game.js*) receives the *start* message and calls the private *#questionNext()* method:

```
// incoming client event
async clientMessage({ type, data }) {

  console.log('Data from client', type, data);

  // handle client event (on single server)
  switch (type) {

    case 'start':
      // fetch first question
```

```
    this.#state.current = type;

    // no question found?
    if (!await this.#questionNext( timerDefault )) {
      await db.broadcast( this.gameId, 'gameover' );
    };
    break;
```

The *#questionNext()* method determines whether more questions can be asked, fetches the next question from the database, and broadcasts it using a *questionactive* message type to all WebSocket servers after a five-second delay:

```
// fetch and broadcast next question
async #questionNext( delay ) {

  // can ask next question?
  if (this.#state.question >= this.cfg.questions_asked) return;

  // fetch next question and answer set
  const qSet = await db.questionFetch( this.#state.question + this.cfg.
  ↪question_offset );
  if (!qSet) return;

  qSet.num = this.#state.question + 1;

  this.#setTimer(async () => {
    await db.broadcast( this.gameId, 'questionactive', qSet );
  }, delay || 1);

  return qSet.num;

}
```

The *questionFetch()* function defined in *Libshared/quizdb.js* returns an object containing the question text and an array of answer objects where one has a *correct* property set to *true* :

```
// fetch next question and answer set
export async function questionFetch( qNum ) {
```

```
// fetch question
const
  qCount = await questionCount(),
  question = await dbSelect('SELECT * FROM question ORDER BY id LIMIT 1
  ↪OFFSET $1', [ qNum % qCount ]);

if (question.length !== 1) return null;

// fetch answers
const answer = await dbSelect('SELECT * FROM answer WHERE question_id=$1 ORDER
↪BY id;', [ question[0].id ]);

if (!answer.length) return null;

return {
  text: question[0].text,
  answer: answer.map( a => { return { text: a.text, correct: a.correct }})
};

}
```

Note that the PostgreSQL *OFFSET* clause fetches the next question according to the random *question_offset* defined for the current game.

Answering a Question

When each client receives the *questionactive* message it runs the *show()* function in the client-side *web/static/js/question.js* script to display the question and possible answer buttons:

```
// CLIENT-SIDE JAVASCRIPT
// show question
export function show( q ) {

  currentQuestion = q;
  currentQuestion.answered = null;

  clear(question);
  clear(answers);
```

```
  answers.classList.remove( answeredClass );

  qNum.textContent = q.num;
  question.innerHTML = q.text;
  currentQuestion.answerNode = [];

  q.answer.forEach((ans, idx) => {
    const button = document.createElement('button');
    button.value = idx;
    button.innerHTML = `<span>${ idx+1 }:</span> ${ ans }`;
    currentQuestion.answerNode[idx] = answers.appendChild(button);
  });

}
```

When the player answers a question—by clicking a button or pressing an associated number (1 to 4) on the keyboard—the *questionAnswered()* function in the client-side *web/static/js/question.js* script verifies it's valid, highlights the button, and raises a custom event named *answered* :

```
// CLIENT-SIDE JAVASCRIPT

// answer event handlers
answers.addEventListener('click', questionAnswered);
window.addEventListener('keydown', questionAnswered);

// user answers a question
function questionAnswered( e ) {

  // already answered?
  if ( !currentQuestion || currentQuestion.answered !== null ) return;

  let ans = null;
  if (e.type == 'click') {

    // button click
    ans = e.target && e.target.nodeName === 'BUTTON' ? parseInt(e.target.value,
    ↳ 10) : null;
    if (ans > currentQuestion.answer.length) ans = null;

  }
```

```
  else {

    // keypress
    ans = e.key >= '1' && e.key <= String(currentQuestion.answer.length) ?
    ↪parseInt(e.key, 10) - 1 : null;

  }

  if (ans === null) return;

  // highlight answer
  currentQuestion.answered = ans;
  answers.classList.add( answeredClass );
  currentQuestion.answerNode[ans].classList.add( answeredClass );

  // raise custom event
  document.dispatchEvent( new CustomEvent('answered', { detail: ans }) );

}
```

This triggers a handler function in the client-side *web/static/js/main.js*
script, which sends a *questionanswered* message to the connected
WebSocket server:

```
// CLIENT-SIDE JAVASCRIPT

// question answered event
document.addEventListener('answered', e => {
  if (state.current === 'questionactive') sendMessage('questionanswered', {
    ↪answer: e.detail });
});
```

This triggers the *Game* object's *clientMessage()* function in *ws/lib/game.js* :

```
// incoming client event
async clientMessage({ type, data }) {

  console.log('Data from client', type, data);

  // handle client event (on single server)
```

```
switch (type) {

  // ...

  case 'questionanswered':
    // player answers question
    if (this.#state.current !== 'questionactive') return;

    // calculate player score
    const correct = this.#state.activeQuestion.answer[ data.answer ].correct;
    data = {
      playerId: data.playerId,
      score: correct ? this.cfg.score_correct : this.cfg.score_incorrect,
      fastest: correct && !this.#state.correctGiven
    };

    // fastest correct bonus?
    if (data.fastest) data.score += this.cfg.score_fastest;

    // first answer controls flow
    if (!this.#state.playersAnswered) {

      let timeout = 100;

      // first response?
      if (!this.#state.playersAnswered && this.player.size > 1) {

        // send question timeout warning
        timeout =this.cfg.timeout_answered * 1000;
        await db.broadcast( this.gameId, 'questiontimeout', { timeout });

      }

      // complete question
      if (timeout) {

        this.#setTimer(async () => {

          // broadcast correct answer
          await db.broadcast( this.gameId, 'questioncomplete', {
            correct: this.#state.activeQuestion.answer.findIndex(a => a.
            ↪correct)
          });
```

```
          // show scoreboard
          this.#setTimer(async () => {
            await db.broadcast( this.gameId, 'scoreboard' );

              // next question or game over?
              if (!(await this.#questionNext( timerDefault ))) {
                await db.broadcast( this.gameId, 'gameover' );
              };

          });

        }, timeout);

      }

    }
    break;

  }

  // broadcast message to all servers
  if (type) await db.broadcast( this.gameId, type, data );

}
```

It calculates the player's score if they're correct, incorrect, or the fastest to respond based on the game settings. This is broadcast to all servers, which update their player scores when they're received by the *#eventHandler()* method (they aren't broadcast to their clients):

```
// incoming event sent to all game servers
async #eventHandler({ gameId, type, data }) {

  console.log('Shared server event', type, data);

  if (gameId !== this.gameId || !type) return;

  // handle server event (on all servers)
  switch (type) {
```

```
    // ...

    // player answers question
    case 'questionanswered':
      if (this.#state.current !== 'questionactive') return;

      const p = this.player.get( data.playerId );
      if (p) {
        p.scoreQuestion = data.score;
        this.#state.correctGiven = data.fastest;
        this.#state.playersAnswered++;
      }

    // ...

  }

  // send to all clients
  if (type) this.#clientSend( type, data );

  // clean up completed game
  if (this.#state.current === 'gameover') {

    db.pubsub.off(`event:${ this.gameId }`, this.#handlerFunction);
    await gameComplete( this.gameId );

  }

}
```

A chain of events then commences on the WebSocket server that received the first answer response:

1   It broadcasts a *questiontimeout* to all servers and clients. When received, each client starts a timer of *game.timeout_answered* seconds, which indicates how long users have to respond (see *web/static/js/main.js*):

```
// CLIENT-SIDE JAVASCRIPT
// receive message
ws.addEventListener('message', e => {
```

```
const { type, data } = parseMessage( e.data );
if (!type || !data) return;

console.log('Data from server:', type, data);

switch (type) {

  case 'questiontimeout':
    startTimer( data.timeout );
    break;
```

2   An identical timer is started on the server. After it has elapsed, it broadcasts a `questioncomplete` message with the correct answer. On receipt, each client runs the `correctAnswer()` function in the client-side `web/static/js/question.js` script to highlight the appropriate button:

```
// CLIENT-SIDE JAVASCRIPT
// receive message
ws.addEventListener('message', e => {

  const { type, data } = parseMessage( e.data );
  if (!type || !data) return;

  console.log('Data from server:', type, data);

  switch (type) {

    case 'questioncomplete':
      question.correctAnswer( data.correct );
      break;
```

3   After another five seconds have elapsed, the server broadcasts a `scoreboard` message to each server. When received, each server appends the calculated player total scores to the message and sends it to its connected clients in the `Game` `#eventHandler()` method (`ws/lib/game.js`):

```
// incoming event sent to all game servers
async #eventHandler({ gameId, type, data }) {

  console.log('Shared server event', type, data);

  if (gameId !== this.gameId || !type) return;

  // handle server event (on all servers)
  switch (type) {

    // show scoreboard
    case 'scoreboard':
      if (this.#state.current !== 'questioncomplete') return;
      this.#state.current = type;
      this.player.forEach(p => p.scoreTotal += p.scoreQuestion);
      data = this.playerAll();
      break;

  }

  // send to all clients
  if (type) this.#clientSend( type, data );
```

On receipt, each client executes the *score()* function in *web/static/js/player.js* to update the player totals. This regenerates the player table with the updated scores with the highest scoring player at the top:

```
// CLIENT-SIDE JAVASCRIPT
// update scores
export function score(pALL) {
  init( pALL.sort((a, b) => b.score - a.score), true );
}
```

4   The server calls the *Game* object's *#questionNext()* method to fetch the next question. This is sent as a new *questionactive* message after another five seconds, and the process restarts.

The method returns *undefined* when the number of questions reaches the *game.questions_asked* . When this occurs, the server broadcasts a *gameover* message to all servers, which is handled by the *#eventHandler()* method in (*ws/lib/game.js*):

```
// incoming event sent to all game servers
async #eventHandler({ gameId, type, data }) {

  console.log('Shared server event', type, data);

  if (gameId !== this.gameId || !type) return;

  // handle server event (on all servers)
  switch (type) {

    // game over
    case 'gameover':
      this.#state.current = type;
      data = {};
      break;
  }

  // send to all clients
  if (type) this.#clientSend( type, data );

  // clean up completed game
  if (this.#state.current === 'gameover') {

    db.pubsub.off(`event:${ this.gameId }`, this.#handlerFunction);
    await gameComplete( this.gameId );

  }

}
```

Each server runs a *gameComplete()* function to delete the *Game* object and the associated record in the database *game* table (only the first will succeed). This causes a cascade of deletions from the *player* and *pubsub* tables for that game:

```
  // remove active game
 async function gameComplete( gameId ) {

  if ( !gameActive.has( gameId ) ) return;

  await db.gameRemove( gameId );
  gameActive.delete( gameId );

  console.log(`Game ${ gameId } removed - active games on this server:
  ↪${ gameActive.size }`);

 }
```

The same *gameover* message is sent to all connected clients. When received, each client shows the **Game over** messages with links to start or join a new game.

Leaving a Game

If the user closes or refreshes their browser, a WebSocket *close* event handler is triggered on the server in *ws/index.js* :

```
 // client connection closed
 socket.on('close', async () => {

   // remove player
   if (player) {
     await player.game.playerRemove( player );
   }

   console.log(`disconnection from ${ req.socket.remoteAddress }`);

 });
```

It calls the *Game* object's *playerRemove()* method in *ws/lib/game.js* :

```
 // remove player from game
 async playerRemove( player ) {
```

```
// delete from database
await db.playerRemove( player.id );

// broadcast event
await db.broadcast(
  this.gameId,
  'playerRemove',
  { id: player.id }
);

}
```

This deletes the player from the database *player* table using the
playerRemove() function in *libshared/quizdb.js* :

```
// remove a player
export async function playerRemove( playerId ) {

  return await dbDelete({
    table: 'player',
    values: { id: playerId }
  });

}
```

It then broadcasts a *playerRemove* message to all WebSocket servers. This is
received by their *Game* *#eventHandler()* , which deletes the player reference:

```
// incoming event sent to all game servers
async #eventHandler({ gameId, type, data }) {

  console.log('Shared server event', type, data);

  if (gameId !== this.gameId || !type) return;

  // handle server event (on all servers)
  switch (type) {

    // remove player
```

```
case 'playerRemove':
  if (this.player.has(data.id)) {
    this.player.delete( data.id );
  }
  break;
```

Finally, the same *playerRemove* message is sent to all clients. On receipt, each client executes the *remove()* function in *web/static/js/player.js* to delete the player from memory and the DOM:

```
// remove existing player
export function remove(p) {

  if (!p.id || !player.has(p.id)) return;

  pList.removeChild( player.get(p.id).node );
  player.delete(p.id);

}
```

Exercises

Try debugging the quiz application using the instructions provided in Chapter 4. It's not as straightforward as before, because a single user could be communicating with any of the HTTP or WebSocket servers.

Fortunately, each player can only connect to one WebSocket server at a time. Examine the Docker log when you start or join a game:

```
ws_1              | connection from ::ffff:172.18.0.3
```

In this case, the user is connecting to the first WebSocket server *ws_1*. Run the following command in another terminal to list the active Docker containers:

```
docker container ls
```

Note the *NAMES* and *PORTS* mappings:

```
PORTS                                                      NAMES
0.0.0.0:59961->8001/tcp, 0.0.0.0:59962->9229/tcp          nodequiz_ws_1
0.0.0.0:59956->8001/tcp, 0.0.0.0:59957->9229/tcp          nodequiz_ws_2
0.0.0.0:59958->8001/tcp, 0.0.0.0:59959->9229/tcp          nodequiz_ws_3
0.0.0.0:59952->8000/tcp, 0.0.0.0:59953->9229/tcp          nodequiz_web_1
0.0.0.0:59954->8000/tcp, 0.0.0.0:59955->9229/tcp          nodequiz_web_2
0.0.0.0:5432->5432/tcp                                    dbserver
0.0.0.0:59951->8080/tcp                                   nodequiz_adminer_1
0.0.0.0:80->80/tcp, 0.0.0.0:8080->8080/tcp                nodequiz_reverse-proxy_1
```

In this example, the following ports are exposed on *nodequiz_ws_1* :

- *localhost:59961* maps to the *ws_1* WebSocket service running on port *8001*
- *localhost:59962* maps to the *ws_1* WebSocket server's debugger running on port *9229*

Open chrome://inspect/#devices in Google Chrome, hit **Configure**, and add *localhost:59962* as a target.

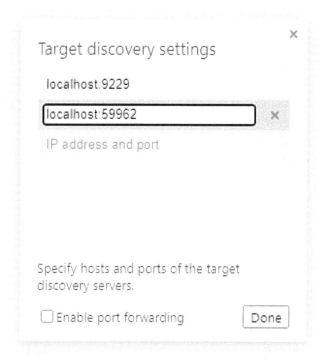

16-2. Adding a debugging port

An *inspect* link to the Remote Target should appear within a few seconds. Click it to open the WebSocket server's debugger.

Next, consider how you could improve the quiz app. For example:

- limit imported questions to specific categories, difficulties, or types
- create administrative screens to add, edit, or remove questions
- allow the user to refresh the page but remain active
- display which players have already answered
- show which choice each player made on the answer screen
- provide "restart game" functionality, which enrolls all current players on a new quiz
- make a game fully recoverable if one or more WebSocket servers fail

Summary

This quiz illustrates how a scalable, multi-server, multi-user, real-time web application can be developed in Node.js using a few third-party modules, vanilla ES6, and less than 60KB of code. Admittedly, negotiating messages between all servers and clients is complex, but that's the nature of multi-player games rather than WebSocket technologies.

In the final chapters, we'll look at a selection of popular Node.js development and deployment tools that you may find useful.

Node.js Tools and Resources

Chapter

17

I hope you now feel confident enough to write your own Node.js programs and find appropriate packages when necessary. The success of the runtime has one downside: *you're spoiled for choice!* There are 1.5 million packages available, ranging from full application development suites to simple, one-function modules. This can lead to choice paralysis, and the moment you settle on one package, a better option will inevitably arrive.

17-1. npm

This chapter provides a list of popular and proven npm packages for use in various situations. They provide a head start, but please don't think you have to use them. Only you can judge whether a package is or isn't useful for your project.

I'll also reiterate a point made throughout this book: *only use third-party modules that are absolutely necessary.* It makes sense to leverage the years of development and real-world testing received by frameworks, database drivers, image optimizers, and so on. You can write smaller modules yourself—such as string or date manipulation functions. It may take longer initially, but should save you time over the long term, because there's no need to search for appropriate packages, manage updates, assess security issues, or switch to alternatives.

Perfect Package Pursuit

The following sites provide curated lists of npm packages:

▩ https://github.com/sindresorhus/awesome-nodejs
▩ https://nodejs.libhunt.com/

You can search for npm packages from the command line. For example:

```
$ npm search mysql

NAME                   | DESCRIPTION        | AUTHOR          | DATE       | VERSION
mysql                  | A node.js dri…     | =felixge…       | 2020-01-23 | 2.18.1
knex                   | A…                 | =tgriesser…     | 2022-03-13 | 1.0.4
sequelize              | Sequelize i…       | =janaameier…    | 2022-02-25 | 6.17.0
mysql2                 | fast mysql driv…   | =sidorares…     | 2021-11-14 | 2.3.3
sails-mysql            | MySQL adapter …    | =particlebanan… | 2021-10-15 | 2.0.0
waterline              | An ORM for Node…   | =particlebanan… | 2021-10-22 | 0.15.0
egg-mysql              | MySQL plugin fo…   | =jtyjty99999…   | 2022-02-11 | 3.1.0
tunnel-ssh             | Easy extendable …  | =agebrock       | 2021-10-03 | 4.1.6
@mysql/xdevapi         | MySQL…             | =ltangvald…     | 2022-01-18 | 8.0.28
hapi-plugin-mysql      | Hapi plugin …      | =adrivanhoudt   | 2022-01-03 | 7.2.6
mysql-abstraction      | Abstraction la…    | =rwky           | 2022-02-22 | 5.1.4
mysqlconnector         | MySQL connector    | =pteyssedre     | 2021-10-26 | 1.0.21
anytv-node-mysql       | Our version…       | =freedom_sherw… | 2022-01-19 | 1.0.0
sql-template-strings   | ES6 tagged templ… | =felixfbecker   | 2016-09-17 | 2.2.2
@keyv/mysql            | MySQL/Mari…        | =jaredwray…     | 2022-01-25 | 1.3.0
aws-xray-sdk-mysql     | AWS X-Ray Patc…    | =aws-sdk-team…  | 2021-11-11 | 3.3.4
winston-mysql          | MySQL transp…      | =charles-zh     | 2021-09-22 | 1.1.1
data-elevator-mysql    | Flexible util…     | =kaasdude…      | 2021-09-29 | 4.0.0
```

An online search engine offers a better interface:

▩ https://www.npmjs.com//a>: the official repository
▩ https://npms.io/: a fast search, which ranks packages by a quality
▩ https://snyk.io/advisor/: ranks packages with a health percentage based on
   their popularity, maintenance, security issues, and contributor community

There are tools for comparing two or more packages:

▩ https://npmcompare.com/
▩ https://moiva.io/

Or tools to extract package information:

- http://npm.anvaka.com/: dependency visualization
- https://npm-stat.com/: download and usage statistics

If you're struggling to choose, opt for a package that:

- is popular
- has a non-restrictive usage license
- receives recent and regular updates
- has a small size
- has the fewest dependencies
- has no major outstanding issues

Most of the packages discussed below satisfy these criteria.

Development Tools

The following packages are tools that aid development rather than form part of your Node.js project. You'll normally install them globally with `npm install <package> -g` or add them as a `devDependency` in the project folder with `npm install <package> --save-dev`:

- https://github.com/nvm-sh/nvm/ (Node Version Manager): manages multiple installations of Node.js
- https://eslint.org/: ESLint finds and fixes JavaScript code problems
- https://www.typescriptlang.org/: Typescript adds variable types and other features to the language and compiles to standard JavaScript
- https://rollupjs.org/: a JavaScript module bundler (a tutorial is available on SitePoint[1])
- https://esbuild.github.io/: a fast module bundler written in Go
- https://postcss.org/: CSS transformer and optimizer (a tutorial is available on SitePoint[2])
- https://www.npmjs.com/package/jsdoc: generates API documentation from JavaScript comments

---

1. https://www.sitepoint.com/rollup-javascript-bundler-introduction/
2. https://www.sitepoint.com/an-introduction-to-postcss/

- https://www.npmjs.com/package/small-static-server: a tiny static file web server
- https://nodemon.io/: restarts Node.js applications when source files change
- https://browsersync.io/: browser live reloads when client-side HTML, CSS, or JavaScript updates

17-2. Nodemon

nodemon has been used throughout this book. Use it in place of *node* when running a script during development to restart the application if a script or any of its modules is changed:

```
nodemon index.js
```

Browsersync is effectively a client-side version of *nodemon* with a few superpowers. The following command starts a web server that can serve HTML files and other assets. Client-side scripts are dynamically reloaded if any *.js* file changes:

```
browser-sync start --server --files "js/*.js"
```

Finally, you'll need a good Node.js-compatible editor such as VS Code[3], Atom[4], or Sublime Text[5]. Most offer extensions for linting, debugging, and

---

[3.] https://code.visualstudio.com/
[4.] https://atom.io/

source-code management.

Testing

Writing tests for your application's internal functions helps ensure updates are robust and won't break existing functionality. Node.js doesn't provide a built-in test runner, but the following packages are popular:

- https://testing-library.com/
- https://jestjs.io/
- https://mochajs.org/
- https://github.com/avajs/ava
- https://github.com/lukeed/uvu
- https://node-tap.org/

The main difference between these packages is the download size and syntax. Most allow you to write English-like assertions, so choose whichever appeals to you or your team.

All suites provide unit testing facilities to verify the result of a function given known inputs. This example uses uvu to test the `Math.sqrt()` method:

```
import { test } from 'uvu';
import * as assert from 'uvu/assert';

test('Math.sqrt()', () => {
  assert.is( Math.sqrt(4), 2 );
  assert.is( Math.sqrt(144), 12 );
  assert.is( Math.sqrt(2), Math.SQRT2 );
});

test.run();
```

The following packages provide headless browser automation tools used for integration testing—that is, testing routes through an application by programmatically clicking buttons and filling in forms to observe an expected

---

5. https://www.sublimetext.com/

result:

- https://pptr.dev/: Puppeteer provides Chrome automation
- https://playwright.dev/: Playwright supports all mainstream browsers
- https://www.cypress.io/: Cypress is commercial option with remote testing

Logging

If you outgrow `console.log()`, third-party logging modules provide more sophisticated logging with messaging levels, verbosity, sorting, file output, profiling, reporting, and more. Popular options include:

- https://www.npmjs.com/package/cabin: Node.js, middleware, and browser logging
- https://www.npmjs.com/package/loglevel: a lightweight Node.js equivalent to the browser `console` API
- https://www.npmjs.com/package/signale: a highly configurable logger
- https://www.npmjs.com/package/pino: a fast and popular Node.js and middleware logger
- https://www.npmjs.com/package/winston: a comprehensive and configurable logger
- https://www.npmjs.com/package/morgan: Express middleware logging
- https://www.npmjs.com/package/storyboard: a logging library that can output to a Chrome DevTools extension
- https://www.npmjs.com/package/tracer: simple log formatting

Full-stack Frameworks

The following frameworks can be used to create full web applications and typically allow rendering on the server, the client, or a mixture, as appropriate. They may offer *hydration* techniques where initial content is generated on the server in HTML before client-side components take over for full interactivity.

17-3. Next.js

- https://nextjs.org/: Next.js is based on React component
- https://nuxtjs.org/: Nuxt.js is based on Vue components
- https://kit.svelte.dev/: SvelteKit is based on Svelte components/li>
- https://sailsjs.com/: Silas is the Node.js equivalent to Ruby on Rails

Server-side Frameworks

If you'd rather have full control over client and server development, the following frameworks primarily handle server-side rendering of HTML content and/or Ajax responses in JSON or any other format:

- https://expressjs.com/: Express is one of the first and most popular frameworks
- https://koajs.com/: Koa is a modern framework designed by the Express team
- https://www.fastify.io/: Fastify claims to be one of the fastest options
- https://hapi.dev/: Hapi focuses on simplicity, security, and scalability
- https://nestjs.com/: NestJS offers concepts similar to the Angular client-side framework

- https://adonisjs.com/: Adonis is the Node.js equivalent to PHP Laravel
- https://feathersjs.com/: Feathers is a lightweight framework for real-time applications and REST APIs
- http://restify.com/: restify is optimized for REST web services

 Node.js in Client-side Frameworks

> Client-side JavaScript (and CSS) frameworks that run in the browser don't generally require Node.js. However, they often use the runtime to provide build tools to scaffold project folders, bundle modules, implement testing, run development servers, or optimize assets at build time.

Web Publishing, Content Management Systems, and Blogging

The following platforms provide administration panels where content editors can write content that's pulled into a site template theme when visitors access the site. These are effectively Node.js alternatives to the PHP-based WordPress.

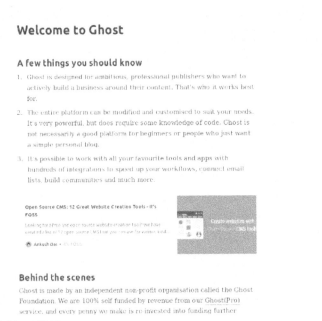

Welcome to Ghost

A few things you should know

1. Ghost is designed for ambitious, professional publishers who want to actively build a business around their content. That's who it works best for.

2. The entire platform can be modified and customised to suit your needs. It's very powerful, but does require some knowledge of code. Ghost is not necessarily a good platform for beginners or people who just want a simple personal blog.

3. It's possible to work with all your favourite tools and apps with hundreds of integrations to speed up your workflows, connect email lists, build communities and much more.

Open Source CMS: 12 Great Website Creation Tools - It's FOSS

Looking for a free and open source website creation tool? We have created a list of 12 open source CMS that you can use for various kinds...

Ankush Das •

Behind the scenes

Ghost is made by an independent non-profit organisation called the Ghost Foundation. We are 100% self funded by revenue from our Ghost(Pro) service, and every penny we make is re-invested into funding further

17-4. Ghost CMS

- https://github.com/TryGhost/Ghost": Gost is a commercial, hosted service is also available at https://ghost.org//a>
- https://hexo.io/: Hexo is closer to an SSG (see the "Static Site Generators" section below), but administration panels can be added via a plugin[6]

Headless Content Management Systems

The following packages provide editing panels and make content available via an API. Articles and other data can be pulled into your application or a static site generator for output to a web page, app, ebook, PDF, or elsewhere:

- https://strapi.io/
- https://keystonejs.com/
- https://apostrophecms.com/

---

6. https://github.com/jaredly/hexo-admin

Static Site Generators

Static site generators (SSGs) pull content—typically from Markdown files or a headless CMS—and place it into templates at build time. The result is a full site pre-rendered as HTML files that can be hosted on any web server without language runtimes, databases, security, or performance implications. You may see this referred to as **Jamstack**[7], which originally meant JavaScript, APIs, and Markup, but is now used in a wider context:

- https://www.11ty.dev/: Eleventy provides Markdown to HTML, with support for many template engines
- https://www.metalsmith.io/: MetalSmith is a simple pluggable SSG
- https://www.gatsbyjs.com/: Gatsby is based on React components
- https://vuepress.vuejs.org/: VuePress is based on Vue components
- https://gridsome.org/: Gridsome is based on Vue components

Database Drivers

Database drivers—also known as **connectors** or **clients**—provide APIs that allow you to connect, query, and update database data. The following packages are *native drivers*: they support a single system, replicate standard commands, and often have official support from the database developers:

- https://www.npmjs.com/package/mysql: for MySQL
- https://www.npmjs.com/package/mysql2: a faster MySQL alternative
- https://www.npmjs.com/package/mariadb: for MariaDB
- https://www.npmjs.com/package/pg: for PostgreSQL
- https://www.npmjs.com/package/mongodb: for MongoDB
- https://www.npmjs.com/package/mssql: for Microsoft SQL Server
- https://www.npmjs.com/package/oracledb: for Oracle
- https://www.npmjs.com/package/couchbase: for Couchbase
- https://www.npmjs.com/package/redis: for Redis
- https://www.npmjs.com/package/sqlite: for SQLite
- https://www.npmjs.com/package/sqlite3: an asynchronous SQLite

---

[7] https://www.sitepoint.com/learn-jamstack/

alternative
- https://www.npmjs.com/package/sqlite-async: a promise-based version of sqlite3

An object-relational mapping (ORM) module can make development easier by providing an abstract layer between your code and the database. Rather than running commands directly, your code manipulates data objects that are saved and restored from a representation in a database. This allows you to switch between systems, but you'll also need to install a native driver, and the full database feature set may not be available. Examples include:

- https://www.npmjs.com/package/mongoose: for MongoDB
- https://www.npmjs.com/package/sequelize: for MySQL, MariaDB, PostgreSQL, SQLite, DB2, and Microsoft SQL Server
- https://www.npmjs.com/package/typeorm: for MySQL, MariaDB, PostgreSQL, SQLite, Oracle, and Microsoft SQL Server

Refer to Chapter 10 for database usage examples.

Templating

Most templating systems generate HTML by inserting values into appropriate blocks. Some provide programming constructs such as file includes, loops, and conditions to optimize development. Popular options include:

- https://ejs.co/
- https://mozilla.github.io/nunjucks/
- https://handlebarsjs.com/
- https://pugjs.org/

Pug differs from others in that you use a concise, indented-style document rather than HTML tags. For example, assume a `title` value is set to "My Site" in the following Pug template:

```
doctype html
```

```
html
  head
    title #{title}
  body
    h1 #{title}
    p#intro Welcome to my site.
```

The resulting HTML is this:

```
<!DOCTYPE html>
<html>
  <head>
    <title>My Site</title>
  </head>
  <body>
    <h1>My Site</h1>
    <p id="intro">Welcome to my site</p>
  </body>
</html>
```

You'll typically use a template system in server-side frameworks such as Express. Chapter 5, Chapter 6 and Chapter 15 of this book use EJS. For example, render an *<h1>* title between a header and footer defined in partials:

```
<%- include('partials/_htmlhead'); -%>

<h1><%= title %></h1>

<%- include('partials/_htmlfoot'); -%>
```

Command Line

The following packages can be useful when creating command-line applications using Node.js:

- https://www.npmjs.com/package/commander: parse command-line arguments
- https://www.npmjs.com/package/cliffy: implement interactive CLIs

- https://www.npmjs.com/package/chalk: output color console messages
- https://www.npmjs.com/package/terminal-link: output clickable hyperlinks
- https://www.npmjs.com/package/boxen: output boxes
- https://www.npmjs.com/package/progress: a simple progress bar

```
progress [=====              ] 29%
```

File System

The standard Node.js library provides an extensive **file system API** for creating, altering, reading, and deleting files and directories. These are fairly low-level functions, so the following packages provide easier file manipulation options:

- https://www.npmjs.com/package/fs-extra: provides a range of file system methods
- https://www.npmjs.com/package/globby: file name string (glob) matching
- https://www.npmjs.com/package/chokidar: cross-platform file watching
- https://www.npmjs.com/package/del and https://www.npmjs.com/package/rimraf: file and directory deletion

Network

The following packages provide a number of network APIs.

Note that a native version of the HTTP Fetch API arrived in Node.js 18. It should become less necessary to use a third-party module as developers and hosts update their installations.

- https://www.npmjs.com/package/node-fetch: HTTP Fetch
- https://www.npmjs.com/package/axios: HTTP Fetch
- https://www.npmjs.com/package/got: HTTP Fetch
- https://www.npmjs.com/package/get-port: get an available TCP port
- https://www.npmjs.com/package/ssh2: SSH client and server methods

WebSockets

WebSockets establish a two-way interactive communication channel between a browser and server, which permits real-time updates and applications. The following packages provide server-side APIs that can send messages to and from the browser WebSocket API:

- <u>https://www.npmjs.com/package/ws</u>: fast lightweight server
- <u>https://www.npmjs.com/package/socket.io</u>: full client and server library

See Chapter 11 and Chapter 16 for WebSocket examples using the `ws` library.

Images

Node.js applications can create, examine, and modify images in most popular formats (JPG, GIF, PNG, etc.) Packages typically provide options to resize, crop, flip, and rotate, or apply filters such as sharpening, blurring, greyscale, and opacity. Popular options include:

- <u>https://www.npmjs.com/package/jimp</u>: scaling, flipping, filters, and pixel analysis
- <u>https://www.npmjs.com/package/image-js</u>: Node.js and browser image manipulation
- <u>https://www.npmjs.com/package/sharp</u>: fast image conversion
- <u>https://www.npmjs.com/package/imagemin</u>: image minification

The following example uses `jimp` to load an image, convert it to greyscale, reduce the width and height by 50%, and output the modified version:

```
import Jimp from 'jimp';

Jimp.read('one.png').then(image => {
  image
    .greyscale()
    .scale(0.5)
    .write('one-bw-small.png');
```

```
});
```

Email

17-5. Nodemailer

The most popular Node.js package for sending email is Nodemailer[8]. The following code sends a single email via an SMTP account:

```
const nodemailer = require('nodemailer');

const transport = nodemailer.createTransport({
  host: 'smtp.example.com',
  port: 587,
  secure: false,
  auth: {
    user: 'username',
    pass: 'password',
  },
});

await transport.sendMail({
  from:     '"Sender" <me@sender.com>',
  to:       'you@recipient.com',
  subject:  'new email',
  text:     'Hello world!',         // plain text body
  html:     '<p>Hello world!</p>',  // HTML body
});
```

An alternative is *node-email* [9], which provides a wrapper around the open-

---

source Sendmail application[10]. Either option is fine for sending ad-hoc emails such as user registration or password reset confirmations.

Bulk email messaging—such as newsletters—is better handled using a dedicated service such as Mailgun[11], SendGrid[12], MailerSend[13], or Mailchimp[14]. These often offer their own Node.js APIs to efficiently manage email transmission.

Finally, `imap-simple` [15] provides a way to connect to and read from an IMAP inbox if you need to provide automated email responses.

Security and Authentication

>Passport[16] is one of the most popular Express-compatible authentication packages for Node.js. It supports more than 500 strategies (plugins) ranging from basic usernames and passwords to passwordless and single-sign-on OAuth options for Google, GitHub, Facebook, Twitter, and LinkedIn.

An alternative option is *grant* [17], which supports more than 200 OAuth providers.

Summary

The Node.js ecosystem is enormous and growing exponentially. Third-party packages are generally designed to handle a single, specific task, so you'll find a range of appropriate options for every situation. The downsides:

---

9. https://www.npmjs.com/package/email
10. https://en.wikipedia.org/wiki/Sendmail
11. https://www.mailgun.com/
12. https://sendgrid.com/
13. https://www.mailersend.com/
14. https://mailchimp.com/
15. https://www.npmjs.com/package/imap-simple
16. https://www.passportjs.org/
17. https://www.npmjs.com/package/grant

- It's easy to become overwhelmed and suffer choice paralysis as you expend time and energy evaluating packages.

- You can become increasingly dependent on third-party solutions. Your development career may descend into writing tedious code to glue packages together.

- The more third-party packages you use, the more time you require to maintain and update that software. You'll often need to update your code as APIs evolve.

There's no such thing as a perfect Node.js package, and I make no apology for repeating my mantra: *only use third-party modules that are absolutely necessary*. Spend most of your time writing code, not choosing tools and resources!

I hope you now have a few simple web projects ready to reveal to the world. The next chapter delves into deployment.

Chapter

Node.js
Application
Deployment

18

You'll eventually want to release your Node.js web app to the world. Deployment options have grown exponentially since the runtime was released in 2009. This chapter describes general types of production hosting, with links to appropriate companies, but the range of services and prices changes daily.

Pages vs Applications

Many readers of this book will be familiar with PHP—the world's most-used web programming language. WordPress alone runs almost half of all websites[1]. A PHP application consists of `.php` files that are interpreted by the PHP runtime when they're accessed via a server such as Apache. HTML or data is then returned to the user's browser.

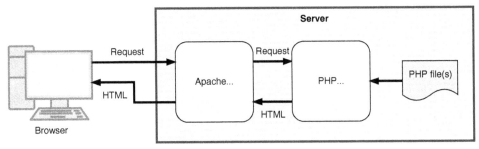

18-1. PHP rendering

The following points are important to note:

- Each page load is stateless. It knows nothing about the application's state, so if a user is logged in, their state must be retrieved from a token or database record during every page request.

- Changing a `.php` file instantly updates the application, because the code is executed when a user requests that resource.

- A `.php` file that causes an error is less likely to cause problems on other pages. Of course, that `.php` file may provide functions shared across multiple pages, but the server and other parts of the application will usually

---

[1] https://w3techs.com/technologies/details/cm-wordpress

remain active.

A Node.js web project is a full application that handles web requests. It doesn't (necessarily) require a server such as Apache and runs continuously *after* the code is loaded from `.js` files.

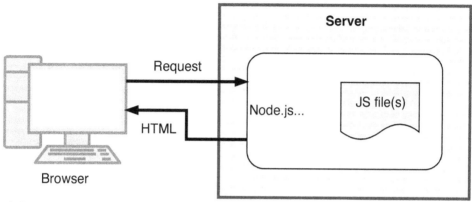

18-2. Node.js rendering

Therefore:

- State *can* be retained. If necessary, a server application could store a JavaScript object for every logged-in user. *(I recommend you write stateless apps, but it's not enforced!)*

- Changing a `.js` file has no impact on the running application. It will only be loaded after the `node` process has been stopped and restarted.

- If any part of your application causes a crash, it goes down for everyone forever! No user will be able to access any part of the system and it will lose any state retained in memory.

The Node.js model has advantages and disadvantages over PHP, but deploying an application to a production server is more challenging.

Most budget shared server hosts support PHP because it can be run by uploading a `.php` file to a server directory. Far fewer offer Node.js, because you require OS-level access to launch an application, which could hog resources as it runs continuously.

Some offer Node.js facilities via systems such as cPanel, where you can define an application's start-up command and configuration. However, these often impose restrictions such as CPU limits or no access to npm.

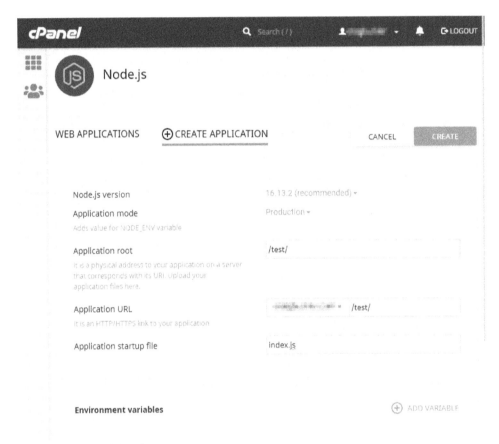

18-3. cPanel Node.js hosting

Node.js Application Preparation

Your development and deployment environments will differ. As a bare minimum, live servers should set the `NODE_ENV` environment variable to `production`:

```
NODE_ENV=production
```

When set, the Express framework disables verbose logging and enables template caching to improve performance. Other modules may offer similar optimizations, but check their README files and documentation.

Internally, your application can detect the `NODE_ENV` value and disable terminal debugging messages, or make other changes such as logging to a file. For example:

```
// running in development mode?
const DEVMODE = (process.env.NODE_ENV !== 'production');

if (DEVMODE) {
  console.log('application started in development mode');
}
else {
  writeToLog('application started in production mode');
}
```

Other environment variables may be required to define application configurations or database connections.

Finally, production servers should normally launch the application with the `node` runtime command rather than `nodemon` or similar. Command-line options such as `--inspect` and `--enable-source-maps` shouldn't be used.

Dedicated Server Hosting

In Node's early days, there was little choice but to spin up a real or virtual Linux server. It requires DevOps personnel to:

1   provision appropriate CPUs, RAM, and disk space

2   install an appropriate version of the Node.js runtime

3   pull the application from a repository

4   `npm install` all project and global dependencies

5   launch the application with `node app.js` as appropriate

Steps 3 to 5 are repeated for every update, although continuous integration and/or continuous deployment solutions can automate the process.

Many hosts offer virtual servers, including DigitalOcean Droplets[2], Amazon EC2[3], Google Compute Engine[4], and Azure Virtual Machines[5].

 sudo-not

> Avoid using `sudo` to run your Node.js application with administrator privileges. The application or any of its modules would have rights to do anything—such as wiping OS files.

HTTP and HTTPS Considerations

Chapter 5 showed how to configure SSL certificates in Express. This isn't recommended on production servers, because the application must be launched using `sudo` to permit use of HTTP ports below 1,000 (port 80 for HTTP or port 443 for HTTPS).

A better option is to launch the application on a non-standard port—such as `3000` —then use a reverse proxy (see the "Use a Reverse Proxy" section below) to forward incoming traffic.

The examples in this book launch development servers on an insecure HTTP connection. This is fine for testing, although care must be taken when referring to internal URLs throughout the frontend and backend code.

---

[2] https://www.digitalocean.com/products/droplets
[3] https://aws.amazon.com/ec2/
[4] https://cloud.google.com/compute
[5] https://azure.microsoft.com/en-gb/services/virtual-machines/

Some developers create a *fake* self-signed certificate for their development server, which makes it more difficult to introduce inconsistent HTTP/HTTPS URLs. The browser will throw an "invalid certificate" error, but you can choose to ignore it. I don't recommend this practice: it has burned me in the past! Browsers often behave differently when they encounter fake certificates. For example, they disable caching, which can lead to strange bugs on live servers.

I recommend you do either of the following:

- Use HTTP during development but be wary when referring to internal URLs that could be HTTPS on production servers.

- Generate a *real*, locally trusted SSL certificate using `mkcert` [6]. These can be used on your own development PC, although you can't share them with other team members (so they'll need to generate their own certificates).

Process Management

Node.js applications run on a single processing thread. In other words, 63 cores in your 64-core server CPU are sitting idle.

You can implement your own clustering code to run an application on all available CPU cores, but this can be difficult (see Chapter 12, as well as the Node.js documentation[7]). A better solution is to use a **process manager**, which makes your live application more efficient by:

- running multiple instances across different CPU cores
- restarting an instance if (*when*) it crashes

For this to work, *your application must be stateless*. Avoid storing application or user state in variables or local files that could differ across instances.

---

6. https://github.com/FiloSottile/mkcert
7. https://nodejs.org/dist/latest/docs/api/cluster.html

18-4. PM2

PM2[8] is the primary contender for Node.js process management. After installing globally, you can start a Node.js application in cluster mode across all available CPU cores:

```
pm2 start app.js -i max
```

 PM2 Port Clashes

> PM2 magically manages ports. If sixteen instances of your Express application all listen on port 3000, PM2 ensures they won't clash. A single request sent to port 3000 gets forwarded to one application instance. The next request may go to another.
>
> Note that PM2 port management can fail if you launch your application using an npm script.

Running processes can be monitored with `pm2 status`.

Use a Reverse Proxy

A reverse proxy passes an incoming request to your Node.js web application. Most web servers can be configured as reverse proxies—including NGINX.

---

8. https://pm2.keymetrics.io/

18-5. NGINX

This has several advantages:

- Any number of domains and applications can be configured on the same server.
- Your Node.js application can be clustered and launched without using *sudo* .
- SSL certificates for HTTPS can be managed by the web server.
- The web server can be configured to serve static assets—such as client-side images, CSS, and JavaScript. This is more efficient than passing the request to Express, because most web servers are multi-threaded.

An NGINX configuration file at */etc/nginx/sites-available/default* can define the incoming ports, set the SSL certificates, look for static files, and resolve requests to the Node.js application when a static file isn't found:

```
server {

  listen 80;
  listen 443 ssl;

  # live domain
  server_name  myapp.com;

  # HTTPS certificates
  ssl_certificate  /etc/nginx/ssl/server.crt;
  ssl_certificate_key /etc/nginx/ssl/server.key;

  # static file?
  location / {
    root /home/node/myapp/static/;
    index index.html;
```

```
    try_files $uri $uri/ @nodejs;
  }

  # Node.js reverse proxy
  location @nodejs {
    proxy_pass http://localhost:3000;
    proxy_http_version 1.1;
    proxy_set_header Upgrade $http_upgrade;
    proxy_set_header Connection 'upgrade';
    proxy_set_header Host $host;
    proxy_cache_bypass $http_upgrade;
  }
}
```

The NGINXConfig configuration tool can help create a setup suitable for your requirements.

Static Site Hosting (Jamstack)

If your application primarily uses client-side HTML, CSS, and JavaScript, it may not be necessary to deploy a Node.js application or use any server-side runtime. A static site generator (SSG) builds directory-based HTML files using content (typically in Markdown format) and templates. There are numerous SSGs[9], but Eleventy[10] is one of the more popular Node.js options.

The resulting build files can be uploaded to any web server. The pages offer:

- excellent performance: they're just files with no server-side processing
- robust security: there's no database or runtime to exploit
- portability: you can host anywhere with no vendor lock-in
- minimum-cost deployments: *often for free*

This simpler approach to web development has become increasingly popular over the past few years. Facilities such as Amazon S3 hosting have been overtaken by platform-as-a-service (PaaS) hosts such as GitHub Pages[11],

---

[9]. https://jamstack.org/generators/
[10]. https://www.11ty.dev/
[11]. https://pages.github.com/

CloudFlare Pages[12], Heroku[13], Vercel[14], and Netlify[15].

Some services offer a simple command-line deployment tool, while others require you to push a branch to a Git repository[16].

 Build PHP-powered Static Sites with Node.js

> I often use Node.js SSGs for smaller websites. These sometimes require basic server-side functionality such as forwarding old URLs or parsing contact forms. Rather than deploy a Node.js server, I output a few `.php` files so the site can be deployed to any PHP host.

Serverless/Lambda Functions

If your app requires more comprehensive server-side processing such as database storage, you could consider serverless functions. Despite the name, **serverless functions** run on a server but there's no need for you to manage the OS, runtime, or even use a framework such as Express.

Serverless functions usually respond on a network endpoint. For example, data posted to https://myapp.com/store-contact/ passes the HTTP request to a function defined in `store-contact.js` , which stores the information and returns a result. The following Netlify serverless function at `functions/hello.js` returns a message when requesting the `/hello/` endpoint:

```
exports.handler = async (event, context) => {
  return {
    statusCode: 200,
    body: 'Hello World'
```

---

12. https://pages.cloudflare.com/
13. https://www.heroku.com/
14. https://vercel.com/
15. https://www.netlify.com/
16. https://www.sitepoint.com/cloudflare-pages-jamstack-deployment/

```
    };
  };
```

You could therefore choose to write a monolithic web application as a series of small stateless functions. These are started on demand, but they usually remain active on busy servers and can scale according to rises in traffic. If a serverless function fails, it's restarted on the next request and won't usually affect or conflict with other functions.

Most cloud hosts offer Node.js serverless functions including AWS, Azure, Google, Cloudflare, Heroku, Vercel, and Netlify.

 AWS Everywhere

> Many serverless hosts, including Netlify and Vercel, deploy serverless functions to AWS Lambda but offer a simpler or improved developer experience.

Serverless functions can be ideal for many applications. They can simplify DevOps and reduce costs on smaller services, but there are downsides:

- **Usage limitations**: not all npm packages can be used, especially if they depend on other OS utilities.
- **Start-up delay**: the first request can take some time as the function is initialized.
- **Shut-down timeout**: functions may have processing limits, so long-running activities such as WebSocket servers may not be possible.
- **Vendor lock-in**: you must adhere to the host's APIs, rules, and updates. It may be difficult to switch to another service.
- **Incalculable costs**: serverless functions are often priced according to compute time. You may have heard anecdotes from developers who deployed a non-terminating recursive function that led to an eye-watering bill.

Container Hosting

You may require more robust hosting as your Node.js service increases in popularity. The multiplayer quiz in Chapter 12 uses Docker containers to launch multiple load-balanced instances of the HTTP and WebSocket applications. The same concept can be applied on production servers. Solutions such as Kubernetes[17] and Docker Swarm[18] can launch, manage, update, and restart containers across any number of servers in any number of locations throughout the world.

At this point, you'll require a dedicated DevOps team to manage deployments costing millions every year. That's unlikely to be a problem: if your app is successful, venture capitalists can't give you enough money, and Google/Microsoft/Apple/Facebook are circling for a potential takeover.

Summary

Node.js hosting options are varied, with extensive ecosystems and prices ranging from free to exorbitant. Personally, I like to write apps that are service agnostic and *could* be hosted anywhere, but that has become more challenging over recent years. We've reached a weird point where you should probably choose a host *before* you write any code. We have numerous hosting solutions, but many companies still select AWS because ... *many companies select AWS!*

Whichever hosting route you choose, you can't go wrong writing stateless web apps. I may have mentioned that a few times before ...

---

[17] https://kubernetes.io/
[18] https://docs.docker.com/engine/swarm/

Chapter

Epilogue

19

Congratulations! You've reached the end of the beginning of your Node.js journey. You've learned a lot, and I hope this course jump-starts your development while helping you avoid some of the pitfalls.

We've covered many topics, from command-line tools, debugging, web applications, and modules, through to real-time, database-driven, multi-player games. No one will fully grasp every topic on their first read, but knowing that a solution exists is half the battle.

I hope you enjoy Node.js development. It has a lot of advantages, such as:

- It's quick to learn the basics and be productive.

- Node.js exposes possibilities you may never have encountered in other runtimes.

- It allows web developers to leverage their client-side JavaScript skills to create useful libraries, frameworks, command-line tools, and even desktop apps.

- Node.js programming can be fun.

Is Node.js for You?

Node.js blossomed from being a niche engine to an indispensable developer runtime within a matter of years. Even those using other languages often have Node.js installed, because it offers a range of tools you won't necessarily find elsewhere.

The reason: *JavaScript*. Web development has become the primary vehicle for platforming applications, so it's difficult to avoid browser-based coding. Using the same language on the frontend and backend lowers the cognitive overhead. Node.js won't make you a full-stack developer overnight, but there's less context switching, and you'll avoid simple errors such as using the wrong quote character, forgetting a semicolon, or making the wrong method call.

Of course, Node.js isn't without its criticisms:

1   Some programmers detest JavaScript.

No language is perfect, but JavaScript was developed in ten days, and it's unlikely Brendan Eich, its inventor, ever considered it might be used for full-scale enterprise level applications. Some issues have been addressed with ES6 and types in JavaScript compilers such as TypeScript.

Personally, I love JavaScript—warts and all. Those who complain loudest are usually comparing it to their favorite language and have been bitten by JavaScript's oddities, such as prototypal inheritance. If it's not to your taste, either persevere or consider one of the many server-side alternatives.

2   npm is cumbersome.

npm is partly responsible for the success of Node.js. It's easy to install, update, and remove any of the 1.5 million packages. Understandably, not every package is *good*, and some have been downright dangerous—laced with malware and crypto-mining code. npm has addressed many issues, but others will occur.

Your `node_modules` directory will also grow to many megabytes and, despite recent optimizations, npm can still recursively download the same packages across different projects. Package maintenance can become increasingly laborious over the long term.

Remember, npm is just a tool. Only install the packages you need and you'll minimize the impact of third-party code.

3   CommonJS vs ES6 module mess.

Node.js is migrating toward ES6 modules, but the process has been painful and some legacy packages may never support it. The situation is improving, though, and I was pleasantly surprised by how few problems I encountered while writing this book.

4   Asynchronous programming is a challenge.

You won't necessarily encounter asynchronous programming in other languages, and it's easy to make mistakes that lead to application instability. I devoted the whole of Chapter 9 to this topic, because it's so important in Node.js programs.

Understanding callbacks can be tricky for novice JavaScript coders, but it's impossible to avoid event handling either on the client or server. Promises and `async` / `await` help, although I initially struggled to understand the concepts.

That said, asynchronous programming makes real-time web applications possible. Instantly updated dashboards, live chat, and multi-player games are far easier in Node.js.

1   Node.js isn't as good/fast/popular/stable/secure as runtimeX.

There will always be alternatives that handle some aspect of application programming in a better way. But Node.js is good enough in most respects for web application and command-line utility development.

To quote C++ designer Bjarne Stroustrup: "There are only two kinds of languages: the ones people complain about and the ones nobody uses."

Is Deno Better?

Ryan Dahl released Deno[1] in 2020 and it addresses many of his Node.js regrets[2]. Deno offers:

- Better security. An application must be granted specific rights when it needs access to environment variables, the file system, the network, and other resources.

[1.] https://deno.land/
[2.] https://www.youtube.com/watch?v=M3BM9TB-8yA

- Native TypeScript support. You can write applications in JavaScript or TypeScript without an additional third-party compiler.

- ES6 modules only. Modules are loaded from a URL: there's no npm equivalent, and packages can be cached so there's one instance on your system across all projects.

- Built-in tools. Linting, formatting, testing, benchmarking, bundling, documentation generation, task running, and more are available from the `deno` runtime.

- Replicated browser APIs. Features such as `window`, `addEventListener`, `Fetch`, and web workers all work in Deno.

- Replicated Node.js APIs. Deno supports features such as `fs`, `events`, `http`, `os`, `process`, `stream`, `url`, `util`, and CommonJS when running in Node.js compatibility mode.

Deno is a great option, but it's new and not as fast, as popular, or well supported as Node.js. Perhaps we'll all be using Deno in a decade's time, and Node.js will be consigned to the history books. But it's too early to tell. There's no harm writing a few small utilities or example apps in Deno … but should you adopt it for a long-term, mission-critical application when it's difficult to find programmers with more than a couple of months' experience?

Deno is similar enough to Node.js that it's easy to switch between the runtimes. Learn Node.js today, then consider Deno tomorrow.

Thank You for Reading!

I hope you enjoyed this book and are ready to embark on the next stage of your programming career. Check out some tips in the final video for this course[3]. Best of luck!

[3.] https://spnt.co/nodevid27

Appendix A: Quiz Answers

Here are the solutions to the quizzes.

Chapter 1

1   d.

2   d. Other than some superficial syntactical similarities, JavaScript has no technical relationship to Java whatsoever!

3   b. TypeScript can compile to JavaScript, but it's a superset of the JavaScript syntax so isn't JavaScript itself!

4   c.

5   a.

Chapter 2

1   d.

2   b.

3   a.

Chapter 3

1   c.

2   d. ... although c. is somewhat extreme!

3   b.

4   a.

5   d. Bonus points if you knew that Docker isn't essential, although it could make Node.js deployments easier!

Chapter 4

1   d.

2   d.

3   b.

4   a.

5   Well, I'm going to say c. It's heavily opinionated, but I don't believe any developer who says they never use `console.log()` ! It's not always the best option and it's too easy to go down a deep console logging rabbit hole, but finding the cause of a bug is more important than the technique you used to get there

Chapter 5

1   d.

2   c.

3   a.

4   c.

5   b.

Chapter 6

1   b.

2   a. But d. *could* be correct if you defined a parsing middleware function!

3   d.

Chapter 7

1   d.

2   b.

3   b.

4   c.

5   d. Bonus points if you realized that a. and b. would list all dependencies in older versions of npm.

6   a.

Chapter 8

1   a.

2   c.

3   a.

4   b.

5   d.

Chapter 9

1   c.

2   d.

3   a.

4   a.

5   d.

Chapter 10

1 b.

2 d.

3 d.

4 a.

5 d.

Chapter 11

1 d.

2 a.

3 d.

4 b.

5 b.

Chapter 12

1 b.

2 d.

3 d.

4 a.

5 b.

www.ingramcontent.com/pod-product-compliance
Lightning Source LLC
Chambersburg PA
CBHW080147060326
40689CB00018B/3881